An Introduction to
Direct Access
Storage Devices

An Introduction to

Direct Access Storage Devices

Hugh M. Sierra

ACADEMIC PRESS, INC.

Harcourt Brace Jovanovich, Publishers

Boston San Diego New York
London Sydney Tokyo Toronto

This book is printed on acid free paper. ∞

ACADEMIC PRESS, INC.
1250 Sixth Avenue, San Diego, CA 92101

United Kingdom Edition published by
ACADEMIC PRESS LIMITED
24–28 Oval Road, London NW1 7DX

Library of Congress Cataloging-in-Publication Data

Sierra, Hugh M.
 An introduction to direct access storage devices/Hugh M.
 p. cm.
 Includes bibliographical references.
 ISBN 0-12-642580-9
 1. Computer storage devices. 2. Data disk drives. 3. Magnetic
recorders and recording.   I. Title.
 TK7895.M4S54   1990
 621.39'76—dc20

89–48406
CIP

Printed in the United States of America
90 91 92 93   9 8 7 6 5 4 3 2 1

This book is dedicated to Patricia, David, Mark, Margaret, and Sandra. With special thanks to my daughter Margaret for typing the manuscript, thus successfully deciphering my scribble.

Contents

Acknowledgments

The first Direct Access Storage Device (Disk Drive RAMAC 350) was conceived and designed by IBM personnel in downtown San Jose, California (99 Notre Dame Street) during the early 1950s. The feasibility models were a hodge-podge of components available at the time: the spindles were obtained from juke boxes, the disks were aluminum pizza plates with a hole in the middle, the magnetic recording materials (liquid) were poured on paper drinking cups and deposited manually, the first read/write heads were assembled by former watchmakers, etc. In fact, some of the early servomotors consisted of coils obtained from hi-fi loudspeakers whose cones had been removed. Today, those motors are still called "voice" coil motors, a term that, although historically correct, is rather anachronistic since there are no "voices" (or music!) to be heard. Also, the coils are designed today especially for disk drives, not for loudspeakers.

Sometimes it is difficult to believe that, from such humble beginnings, a $20 billion dollar international industry has emerged in such a short period of time. In a very large measure, this has been due to the enthusiasm, dedication, and intelligence of many individuals who have contributed steadily to the development and improvement of these devices. So many people have contributed that it would be impossible to acknowledge them all.

The author considers himself very fortunate indeed to have been able to meet most of these people personally during a period of nearly 35 years, some of them as co-workers, later on as colleagues, and finally as personal friends. This was greatly facilitated by the geographical coincidence that a great number of those advances took place in High Technology Valley (a.k.a., Silicon Valley, an appellative that, although rather quaint and catchy, happens to be wrong, and if nothing else, is an unwarranted exaggeration).

Preface

The objective of this book is to explain the organization and internal structure of present-day Direct Access Storage Devices (disk drives or DASDs). The book has been written primarily, but not exclusively, for the following:

(1) Computer Scientists (students and/or professionals), who by understanding the arrangement and functioning of DASDs, should be able to circumvent their limitations and take better advantage of their strengths.
(2) Engineers (students and/or professionals) who will find DASDs an extremely challenging application of their skills.

The presentation assumes a minimum level of knowledge in magnetic recording, servomechanism design, and coding. However, this is not indispensable because these subjects are treated here in just the amount required. Those interested in a more in-depth treatment are referred to the Bibliography. While all the data have been checked for accuracy, it is possible that by the time you read this book, changes and/or improvements may have occurred. Nevertheless, this book contains much valuable information. Although DASDs are an extremely vast subject, only the most pertinent portions have been included, and the mathematical treatment includes only those equations needed to ensure clarity. Any errors of omission or commission may be blamed entirely on the author.

For the most part, the organization of the book is such that the chapters follow the path of a readback signal. That is, the digital pulse originates in a disk platter, it is picked up by a transducer (the Read/Write head), amplified, modified, and presented to the interface, ready to be used by a computer. Specifically, Chapter 1 offers the trends in implementation and usage of DASDs. Chapter 2 constitutes a brief introduction to magnetic recording, the method of preference today for the storage of information. Chapter 3 is a compilation of formulas used during the proposal and/or evaluation of a DASD. Chapter 4 discusses one of the methods that has been used to increase the capacity of DASDs. Chapter 5 is an exposition of the interaction between the Read/Write transducers and the revolving platters. Chapter 6 is an enumeration of the different methods that have been used to encode the digital data to be stored in a platter. Chapter 7 covers in detail one of these codes, namely, the RLL(2, 7). Chapter 8 explains the circuitry required to store and retrieve the digital pulses. Chapter 9 consists of a brief introduction to the servomechanism needed to place the Read/Write heads on the desired location of the platter (track). Chapter 10 is a preliminary mathematical comparison between magnetic and optical recording, as used in DASDs. Chapter 11 offers one of the many methods that can be used at the system (or programming) level to exploit the advantages offered by DASDs.

It is impossible to expect every reader to be interested deeply in each and every one of the subjects covered in this book. Nevertheless, a great deal of effort has been spent in making every subject self-contained. Consequently, those items of no particular interest can be skipped over without any loss of continuity.

A word of caution: Many items in the Bibliography consist of Patents which are still active. Therefore, although the subject matter might have been amply documented (but not in a single place), this book does not constitute a "license to kill." The prospective user will be well advised to check the appropriate licenses before trying to implement some of the techniques described here, either at the component or system level.

Notations and Abbreviations

S = Rotational speed in revolutions per second (RPS); constant for all tracks

t_r = Period of one revolution; constant for all tracks

Δ_B = Data rate in bits per second at the interface to the controller; constant for all tracks

Δ_F = Data rate in flux changes per second at the R/W head; constant for all tracks

N_B = Number of bits per track at the interface to the controller; constant for all tracks

N_{BY} = Number of bytes per track at the interface to the controller; constant for all tracks

N_F = Number of flux transitions per track at the R/W head; constant for all tracks

ℓ = Track number; also used as a subscript

I = Innermost track; used only as subscript

O = Outermost track; used only as a subscript

$T = TPI$ = Track density in tracks per inch

$B = BPI$ = Bit density in bits per inch

A = Areal density in bits per square inch

C_B = Unformatted capacity per disk side in bits (data bits)

C_{BY} = Unformatted capacity per disk side in bytes (data bytes)

C_T = Unformatted total capacity per drive in bits (data bits)

C_{TY} = Unformatted total capacity per drive in bytes (data bytes)

DS = Number of data surfaces

D = Track diameter in inches

R = Track radius in inches

G = Track linear length

K = Total number of tracks per disk side; compression ratio

Y = Latency

H = Radial width of data band

θ = Angle

$P(x)$ = Probability of event x occurring.

$\lambda(t)$ = Failure rate

λ = Constant failure rate

m = True MTBF; MTBF requirement MTBF prediction

p = Probability of success

$R(t)$ = Reliability function; cumulative probability of survival

R_s = System reliability

q = Probability of failure

P_T = Track pitch

DR = Density ratio of the magnetic recording code used

FR = Frequency ratio of the magnetic recording code used

d = Minimum number of consecutive zeros allowed (including clock)

k = Maximum number of consecutive zeros permitted (including clock)

m = Minimum number of data bits to be encoded

n = Number of code bits (including clock) for each of the m data bits

r = Number of different word lengths in a variable length code

w = Detection window expressed as a percentage of a data bit cell

1F = "Alternate Ones" fundamental frequency at the R/W head

2F = "All Ones" fundamental frequency at the R/W head

F_L = Lowest frequency at the R/W head

F_H = Highest frequency at the R/W head

T_{FL} = Half-period of the lowest frequency at the R/W head

T_{FH} = Half-period of the highest frequency at the R/W head

W_F = Spacing, in seconds, between flux changes

T_{WF} = Half-space, in seconds, between flux changes

T_W = Half-space, in seconds, between bits at the controller interface = half data window

W_{FB} = Spacing, in seconds, between bits at the controller interface = data window

T_C = Data cell = data window = W_{FB}

L = Linear speed (of some part of the disk) in inches per second

W_i = Input pulse width at the base

W_o = Output pulse width at the base

Tables

1

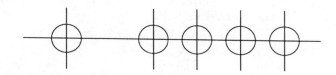

Trends in
Direct Access
Storage Devices

After many years of development, certain trends have become clearly discernible during the usage and implementation of Direct Access Storage Devices (DASDs). The trends for the most important parameters can be seen by plotting the known data on semi-logarithmic paper. These graphs serve not only as historical documents but also as indicators of the quantities predicted for the near future. This chapter also explains the basic philosophy behind the requirements needed to continue this increasing capacity of DASDs. In addition to some implementation parameters, some marketing trends have become apparent and are included here.

1.0. INTRODUCTION

Since their inception in the early 1950s, disk drives have undergone significant improvements. For example:

(A) Linear recording density B has gone from 100 bits per inch to 32,000.

Table 1.1.

Disk Drive Improvements Expected within a Few Short Years.

Parameter	Improvement
Access Time	Reduced by a factor of 1.5
Latency (Rotational Delay)	Reduced by a factor of 2
Power consumption	Reduced by a factor of 1.5
Weight	Reduced by a factor of 1.7
Cost	Reduced by a factor of 8
Capacity	Increased by a factor of 10
Disk drive reliability	Increased by a factor of 2

(B) Track density T has gone from 20 tracks per inch to 1,500.

(C) Cost per megabyte has been reduced from \$25,400 to \$10.

(D) Disk platter diameters have been diminished from 40 inches to $3\frac{1}{2}$ inches.

(E) Mean time between failures (MTBF) has increased from a few hundred hours to 30,000 hours.

Also:

(A) Disk drive capacity has doubled every four years.

(B) Market demand has doubled every two years.

Consequently, it is safe to estimate that this trend will continue for the foreseeable future. Table 1.1. shows several improvements in disk drives that we can expect within a few short years.

1.1. CONSTRUCTION

In spite of these improvements, the basic construction has remained unchanged throughout those years. Even today, rigid disk drives consist of five basic sub-systems:

(1) Spindle (with rotational motor and bearings),

(2) Circular disks (aluminum substrate covered with some magnetic compound),

(3) Read/Write heads (with sliders and suspensions),

(4) Electronic circuitry (Read/Write channels, servocontrol, power supply, interface, etc.), and

(5) Servomechanism.

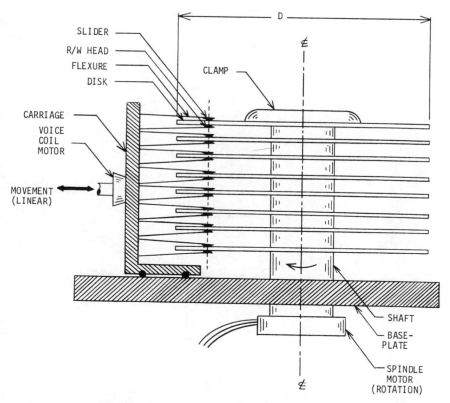

Fig. 1.1. Implementation of a multi-disk rigid drive.

Figure 1.1. shows a block diagram of a rigid disk drive: a vertical spindle is held in place by bearings and it is made to rotate at a very uniform speed (accurately controlled). The spindle holds in place several rather thin circular platters, usually made of some aluminum alloy and on whose lower and/or upper surface(s) a magnetic material has been deposited.

At the same time, there are a number or Read/Write (R/W) heads (one for each available disk surface). Since all the R/W heads are in a vertical position, they comprise a cylinder. These heads are mounted on specially designed sliders (which keep them in close contact with their respective disk surface) and held at length by flexures. The whole ensemble of R/W heads, sliders, and flexures are attached to a carriage that is made to move in or out of the disks in a radial direction. The motive force for this movement is provided by a "voice" coil motor controlled by a servomechanism.

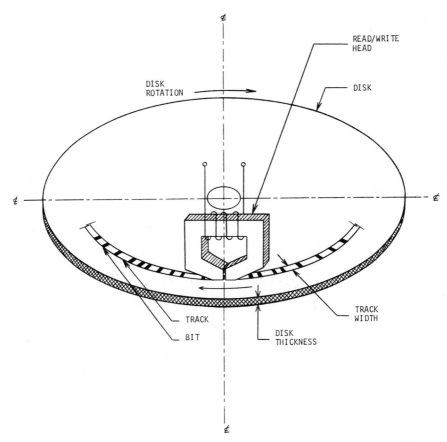

Fig. 1.2. View of a single head and a single disk. (1) While reading or writing the R/W head remains stationary. (2) When moving from track to track the R/W head moves in the radial direction of the disk.

For purposes of explanation, Fig. 1.2. shows a single R/W head and disk. While reading or writing on the disk surface, the R/W head remains stationary. However, when the servomechanism circuitry receives an instruction from the controller, the whole carriage is made to move in a radial direction until the desired location is reached. The unique ability of these drives to reach any desired location in a random fashion, in a relatively short period of time, and at low cost, makes them very attractive for storing large amounts of data in a data processing environment.

Since the R/W heads remain stationary while the data is being read or written, and since at the same time the disks keep rotating underneath,

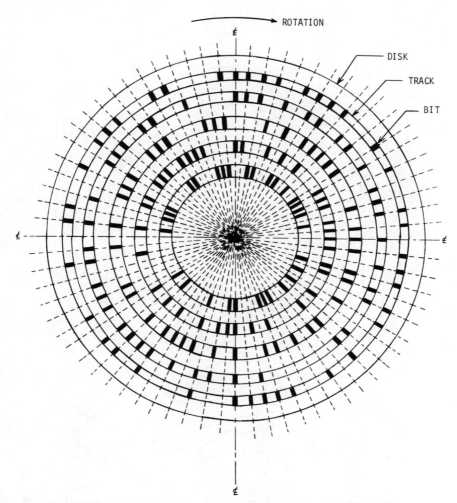

Fig. 1.3. Arrangement of the tracks on a disk as concentric circles. To maintain a constant data rate, all the tracks contain the same number of bits, regardless of their diameter.

the end result is that the data on the disk surfaces consist of concentric circles called tracks. This is unlike hi-fi records in which the data (speech or music) consists of a continuous spiral. Instead, in disk drives, the tracks consist of concentric circles, each basically independent of the others as shown in Fig. 1.3. For purposes of clarity, only a few tracks and bits have been illustrated, although, a disk drive consists of several thousand tracks and several million bits.

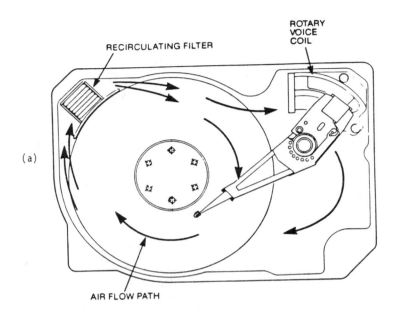

(a)

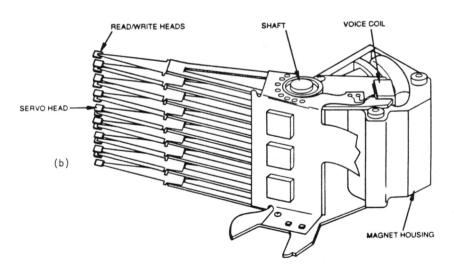

(b)

Fig. 1.4. Packaging of a commercially available disk drive. (a) Disks and R/W heads. (b) Heads, suspensions, and voice coil actuator. © 1989, Maxtor Corporation.

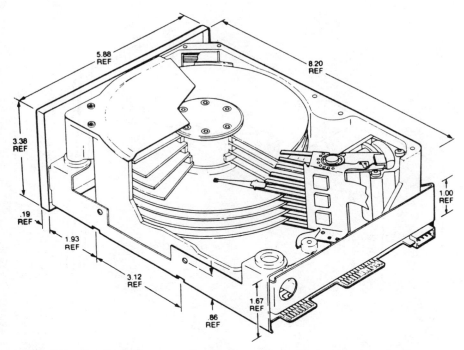

Fig. 1.5. Cut-away view of a commercially available disk drive showing the spindle, disks, heads, suspensions, and voice coil actuator. © 1989, Maxtor Corporation.

Obviously, all of these subsystems are contained in a suitable package. Figures 1.4., 1.5. and 1.6. display a commercially available disk drive similar to the ones available today in the marketplace.

1.2. CAPACITY

Figure 1.7. is a semi-logarithmic graph of disk drive capacity through the years. In 1957, the capacity was only five megabytes, while today this parameter has become 7,500 megabytes. Two observations seem pertinent:

(1.) Except for the sudden drop that took place in 1964, the capacity curve is very smooth and ascending, indicative of a technology in which the increases are made in rather small but relentless steps. The discontinuity in 1964 happened when it was decided that disk packs be made portable, and at that time, this was only possible by reducing the number of data surfaces (DS) substantially, thus diminishing the disk drive capacity.

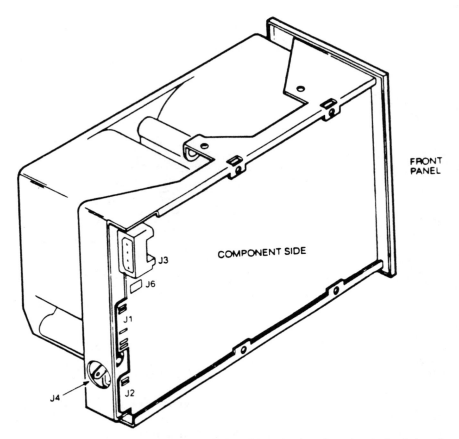

Fig. 1.6. Commercially available rigid disk drive showing the printed circuit board containing the electronic circuitry. © 1989, Maxtor Corporation.

(2) Every one of the models shown in that curve had to offer some improvement over previous models. Even the IBM 3340 (Winchester) was such an improvement. It was rather preposterous to label a whole technology by the code name (Winchester) of a single model (the IBM 3340), which had a first customer shipment (FCS) in the year 1973, because this misnomer ignored the previous 16 years of development that made the Winchester possible as well as the following 16 years of similar improvements!

The increases in disk drive capacities have been in large part due to many improvements in technology. Of course, this has been neither easy

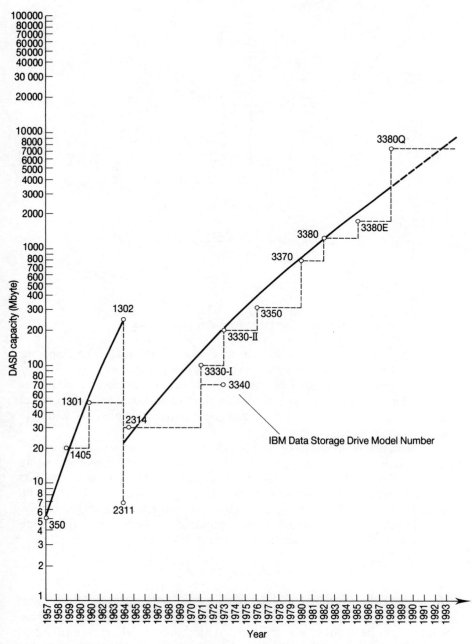

Fig. 1.7. Disk drives capacity in megabytes vs. year of First Customer Shipment (FCS). IBM machines only.

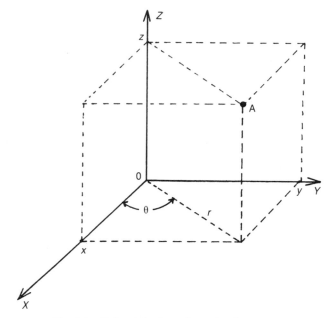

Fig. 1.8.   Point *A* in three-dimensional space.

nor straightforward. It has taken a tremendous amount of development effort and money, and has involved quite a few disciplines.

Let us call *B* the number of bits per inch in the innermost track. Let us call *T* the number of tracks per inch in the radial direction. The product *B* × *T* gives the areal density in bits per square inch possible with the technology at hand.

However, this product *B* × *T*, though used for comparison purposes, is rather fictitious. This is because disk drives are constructed by using circular disks and, as shown in Fig. 1.3., the magnetically recorded bits are written in a circular pattern, not in a square pattern. This can be visualized also by considering a geometrical abstraction, as shown in Fig. 1.8. In Cartesian coordinates, the point *A* is defined as the three dimensions of a *cube* for which the point *A* is a vertix:

$$A(x, y, z).$$

However, it is not mandatory to use a cube. It is possible also to use another geometrical figure, such as a cylinder. In this particular case, the point *A* can be defined also by three dimensions in which one of them happens to be an angle (Θ). Therefore, in cylindrical coordinates, the

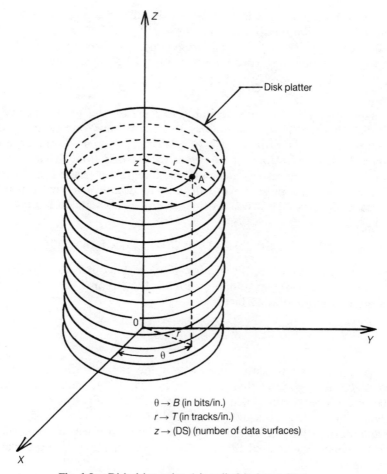

Fig. 1.9.   Disk drive point A in cylindrical coordinates.

same point A can be defined as follows:

$$A(\Theta, r, z).$$

A disk drive can be considered a cylinder, consequently, any bit A in any one of its data surfaces (DS) can be defined by three coordinates, as shown in Fig. 1.9.

In a disk drive, the coordinate Θ can be associated with the linear recording density B in bits/inch. The coordinate r can be associated with the linear track density T in tracks/inch. Finally, the coordinate z can be associated with the number of data surfaces (DS).

The total capacity C_{TY} of a disk drive in bytes is directly proportional to the *product* of the three coordinates mentioned above. Mathematically:

$$C_{TY} \propto (B) \times (T) \times (DS).$$

Consequently, if it becomes desirable to increase the total capacity C_{TY} of a disk drive, it is not mandatory to increase any one of those three coordinates *alone*. On the contrary, it is only necessary to increase each of them by a *smaller* amount to obtain the desired total. That is, in this case, we are not interested in each of the factors but in the final product.

For example, suppose it were desired to double (factor of two) the total capacity of a disk drive. In this case, it is possible to increase *each* factor by a mere 26% because

$$(1 + 0.26) \times (1 + 0.26) \times (1 + 0.26) = 1.26 \times 1.26 \times 1.26 = 2.$$

This can be summarized by stating that in 1957 the total disk drive capacity was 5 megabytes, while by 1988 it has become 7,500 megabytes, thus offering an increasing factor of

$$\frac{7,500}{5} = 1,500.$$

It is very tempting to state that without the capacity discontinuity that took place in 1964 (and that is shown in Fig. 1.7.), today's capacity of 7,500 megabytes could have been obtained sooner (around 1973) as shown in Fig. 1.7. Such a statement would be false because it is based on the assumption that the original number of disk surfaces and platter size have remained unchanged throughout the years. However, this has not been the case.

First, let us concentrate exclusively on magnetic recording technology and assume that the number of data surface (DS) has not changed at all although the number of data surfaces has actually decreased considerably. The linear magnetic recording density B has increased by a factor of

$$\frac{32,000}{100} = 320,$$

while the linear track density T has increased by a factor of

$$\frac{1,300}{20} = 65.$$

It seems easy to say that the capacity should have increased by a factor of

$$(320) \times (65) = 20{,}800$$

instead of only 1,500.

However, we must take into consideration the reduction in platter disk size that occurred during this time.

The disk drive platters used in 1957 had a diameter of 36 inches. The area of a 36-inch diameter circle is:

$$A = \left(\frac{\pi}{4}\right) \times (36)^2 = 1{,}018 \text{ square inches.}$$

Therefore, when B and T were extremely small (by today's standards), it became imperative to use large diameter platters that had large areas and acceptable drive capacities. However, when platter portability became desirable, the platters started to become smaller. For example, by 1964, the platters had a diameter of 24 inches with an area reduction factor of

$$\left(\frac{36}{24}\right)^2 = 2.25.$$

Today, the IBM 3380 has a platter 14 inches in diameter, which gives an area reduction factor of

$$\left(\frac{36}{14}\right)^2 = 6.61.$$

Consequently, while the increasing magnetic recording density was making possible an increase in capacity, the smaller platter diameters were reducing those theoretically possible capacity increases.

Nevertheless, in spite of those two opposing constituents, the drive capacities kept increasing, as shown in Fig. 1.7., although not by factors as large as 20,800.

For example, because of the area reduction of 6.61 calculated above, the capacity increase factor is only

$$\frac{20{,}800}{6.61} = 3{,}147,$$

and at the same time, if we consider that the number of data surfaces (DS) has been reduced by an approximate factor of two, we obtain an increasing capacity factor of 1,500 as being offered today.

The trend to keep reducing the platter diameters (with the accompanying reduction in drive capacity) continues unabated simply because, at the magnetic recording densities being considered today, heat expansion, vibration, etc. become problematic.

Table 1.2.
Expected Demise of Disk Drive
Platter Sizes. Note: The Disk Sizes
40″, 36″, and 24″ are Already Dead.

Platter size	Year
14″	1989
8″	1991
$5\frac{1}{4}$″	1993
$3\frac{1}{2}$″	?

Table 1.2. shows the expected year of demise for several commonly used platter diameter sizes.

1.3. FIGURE OF MERIT

The disk drive capacity is usually dictated by data processing requirements. However, there is a figure of merit that is dictated by economical reasons: This is the cost per megabyte, expressed by the ratio

$$\frac{\$}{\text{megabyte}} . \tag{1.1}$$

Because disk drives must compete with other technologies, such as tape drives, semi-conductor memories, bubbles, thin-film storage, core memories, etc., the cost per megabyte of data stored in disk drives has been decreasing steadily through the years. Table 1.3. shows the cost per megabyte for several disk platter diameters, both past (1985) and future (1989).

Table 1.3.
Cost per Megabyte for Several Disk Platter
Diameters.

Disk Diameter (Inches)	$/Megabyte	
	Year 1985	Year 1989
$3\frac{1}{2}$	—	1.5
$5\frac{1}{4}$	7.5	3.5
8	10	5
14	12	6.5

Equation (1.1) above can be decreased using three methods:

(1) By decreasing the numerator,
(2) By increasing the denominator, or
(3) By a combination of the two methods above, as long as the numerator decreases by a smaller factor than the denominator increases.

Example. Let us consider the case when the cost per megabyte must be cut in half.

Method 1: Cut the cost of the disk drive in half but leave the capacity untouched ($1/2 \div 1 = 1/2$).

Method 2: Charge the same price for the disk drive (keep the cost the same) but double the capacity ($1 \div 2 = 1/2$).

Method 3: Double the cost of the disk drive and, at the same time, quadruple the disk drive capacity ($2 \div 4 = 1/2$).

1.4. MARKETING

Marketing is a very extensive subject and warrants several books. Because disk drives have become so pervasive and plentiful, they have been divided into three categories according to the technology used. These categories are the following

(A) Rigid, magnetic,
(B) Floppy, magnet,
(C) Optical recording (both rigid and floppy).

These three categories have been further subdivided according to capacity. This results in a plethora of sub-categories whose study becomes quite lengthy. In this introductory book we are going to illustrate only one of these sub-categories, namely, rigid disk drives with more than 500 megabytes capacity. The interested reader is encouraged to consult the literature on the complete subject (Porter, 1987).

At this moment, the list of rigid disk drive manufacturers include the following:

38	U.S. Manufacturers
21	Asian Manufacturers (mostly Japanese)
11	European Manufacturers
Total: 70	worldwide

The field of rigid disk drive manufacturing is apparently well populated and competitive.

However, when we consider the sub-category of large capacity rigid disk drives (larger than 500 megabytes) alone, the field narrows considerably and includes only a few manufacturers that offer platters measuring 14 inches, 10 inches, 8 inches, and $5\frac{1}{4}$ inches in diameter.

Figure 1.10. shows a semilogarithmic plot of the past, present, and projected revenues produced by rigid disk drives with more than 500 megabytes of capacity. These revenues have been divided into three categories:

(a) CAPTIVE. These are the drives that some manufacturers use for their own data processing products. They are not offered for sale to the general public unless they are bought attached to a computer made by the same manufacturer.
(b) PCM (Plug Compatible Manufacturer). These are the disk drives that compete directly with captive drives and are designed to displace them by being "plug-compatible" with them.
(c) OEM (Original Equipment Manufacturer). These are disk drives offered to all users to be incorporated in non-captive, non-PCM environments.

Since the total market consists of an ascending curve, as shown in Fig. 1.10., it is apparent that these drives produce a tremendous amount of revenue. For example, during 1987, these drives produced more than $10 billion worth of revenue for their manufacturers.

It is interesting to find out how the market is divided amongst the three categories mentioned above. For this purpose, let us call 100% the total revenue at a given time and calculate the percentage that corresponds to each of these categories. The graph that results is shown in Fig. 1.11. in a semilogarithmic paper also. It is seen that both the captive and PCM markets are shrinking substantially, while the OEM portion is increasing at a great pace.

Also, for purposes of illustration, let us view the market segmentation at a point in time, say 3Q87 (third quarter of the year 1987). The result can be seen in Fig. 1.12 in pie-chart form.

1.5. SUMMARY

This chapter has presented, in general terms, the construction of disk drives plus the trends that have become apparent during their usage, design, and implementation. These trends not only place the devices in

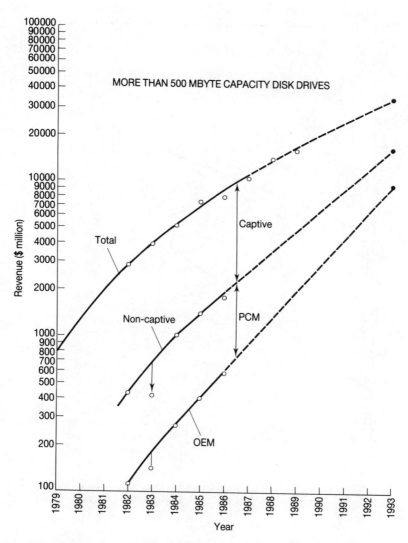

Fig. 1.10. Past, present, and projected revenues for rigid disk drives with more than 500 megabytes capacity.

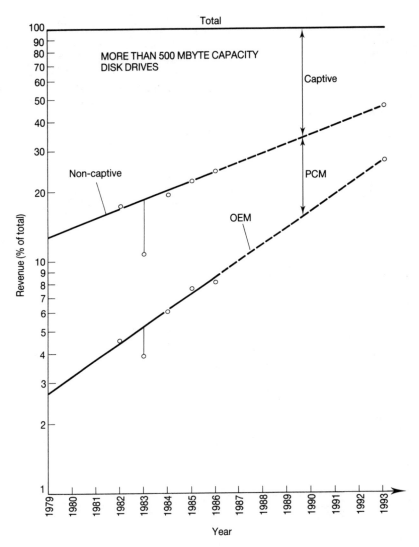

Fig. 1.11. Past, present, and projected revenues for rigid disk drives with more than 500 megabytes capacity as a percentage of total.

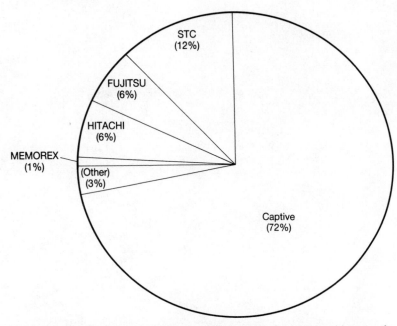

Fig. 1.12. Market segmentation during 3Q87 for rigid disk drives with more than 500 megabytes capacity.

the present time but also offer insight into their immediate future. Consequently, these trends become quite useful to design engineers and to computer scientists responsible for the inclusion of DASDs in a data processing environment.

2

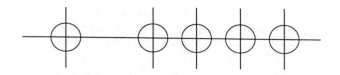

Magnetic Recording for Direct Access Storage Devices

A very large number of excellent books and articles exist on the general subject of magnetic recording for analog and digital applications, some of which have been pointed out in the Bibliography. Furthermore, at the present time, magnetic recording seems to remain the method of preference for the storage of digital data. However, this technology is so vast that in this introductory exposition, it becomes mandatory to reduce this technology considerably by covering (very briefly) just a few applicable topics, such as linear pulse superposition, write pre-compensation, equalization, and horizontal versus vertical digital recording. The reader interested in a thorough knowledge of this subject is encouraged to consult the Bibliography. This chapter constitutes an introduction to some pertinent subjects in magnetic recording that influence the performance of DASDs and the methods employed during their implementation. Because there are still many limitations to the magnetic recording approach however, alternative methods of recording continue to be investigated.

2.0. MAGNETIC RECORDING

Most present disk drives are based on magnetic recording technology. After more than 40 years of development (for saturation recording—for bias recording, even longer), this technology has become very stable and predictable. For example, for rigid disk drives, Fig. 2.1. shows the magnetic recording linear density B in bits/inch versus year of first customer shipment (FCS) for IBM machines.

It can be observed that in a semilogarithmic graph the plot becomes and ascending straight line. Therefore, if we assume that this trend will continue for the forseeable future (not a bad assumption altogether!) it becomes possible to ascertain rather accurately the position in time of any disk drive technology. In 1957, B had a value of 100 bits/inch while by 1988 this parameter was 32,000 bits/inch. Figure 2.1. seems to indicate that magnetic recording technology has settled into a comfortably ascending straight line we can expect will continue in the near future.

Likewise Fig. 2.2. shows in a similar semilogarithmic graph the linear track density T (in the radial direction of the disk) in tracks/inch versus years of FCS for IBM machines. In 1957, T had a value of 20 tracks/inch while by 1988 this parameter was 1,300 tracks/inch.

The product $B \times T$ gives the possible areal density A in bits/square inch. In 1957, this product was 2,000 bits/square inch, while by 1988 it was 41.6 million bits/square inch. This is shown in Fig. 2.3.

The three plots mentioned above result in ascending straight lines. Because these lines ascend as time goes on, magnetic recording has been labeled a "moving target."

Those graphs not only give a "bird's eye" view of the whole field, but they can be used also to evaluate the offerings of any disk drive manufacturer (to ascertain where they are) or any new proposal (as done in Chapter 3). At the same time, they become quite useful when comparing any technology to magnetic recording (as in Chapter 10).

2.1. PROCESSES

Let us divide magnetic recording technology into two distinct processes: Write and Read.

Write: The whole phenomena of writing on a magnetic surface consists of changing the magnetic polarity from north to south or from south to north, as the case may be. This is effected by the write driver according to the instructions received from the controller. Since it is impossible to build up a current in zero time, and since at the same time the disk keeps

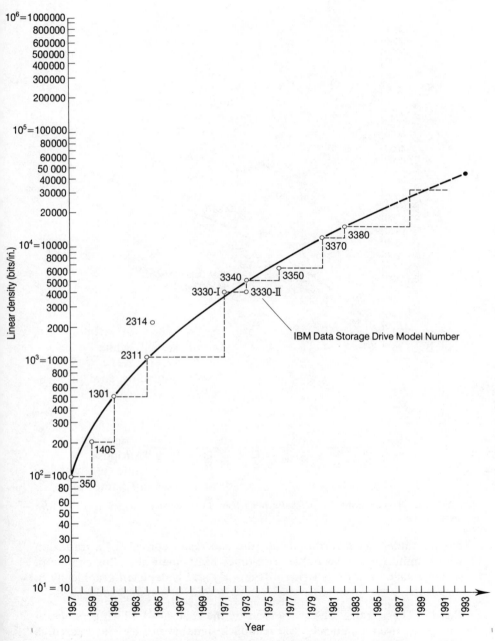

Fig. 2.1. Linear density B in bits/inch versus year of first customer shipment (FCS). IBM machines only.

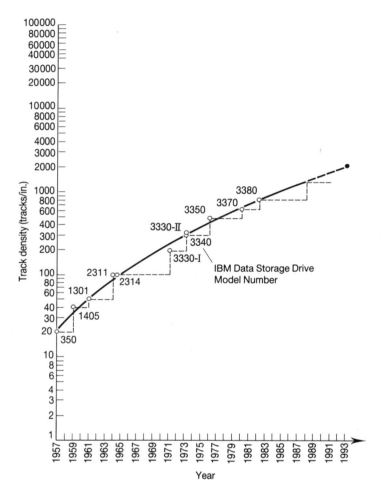

Fig. 2.2.   Track density T in tracks/inch vs. year of first customer shipment (FCS). IBM machines only.

moving under the recording head, the end result consists of a transition with a finite (and measurable) rise-time. Mathematically, this change in surface magnetization is written $M(x - x_0)$ and is depicted graphically in Fig. 2.4.

Read: These transitions are sensed by the read head which produces a voltage e_0. Mathematically, the readback signal (voltage) is expressed as follows:

$$e_0 = CV \frac{\partial}{\partial x} \int_{-\infty}^{\infty} D(x)M(x - x_0)\,\mathrm{d}x \qquad (2.1)$$

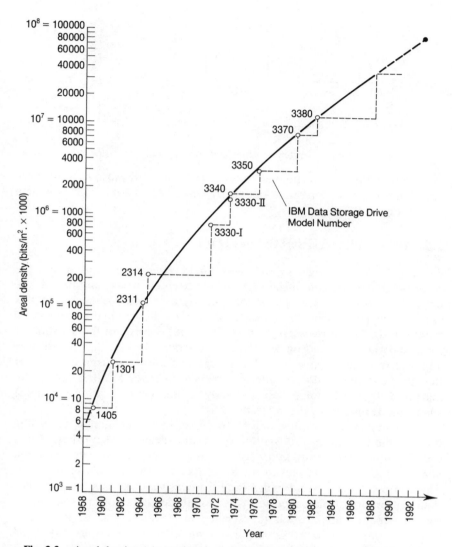

Fig. 2.3. Areal density A in bits/square inch vs. year of first customer shipment (FCS). IBM machines only.

where $D(x)$ represents the sensitivity function of the R/W head and, as indicated above, $M(x - x_0)$ is the change in surface magnetization. The integration is nothing more than a convolution, therefore, the sensitivity function can be considered a filter that degrades $M(x - x)_0$ during the readback process. Consequently, $M(x - x_0)$ is narrower (in time and space) than $D(x)$. At the same time, this convolution is differentiated, as

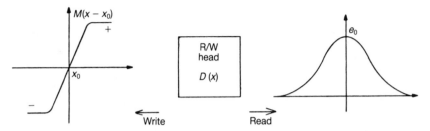

Fig. 2.4. Isolated written transition and corresponding read-back pulse.

indicated in Eq. (2.1). The end result is that the read pulse e_0 looks very much like the first derivative of the transition $M(x - x_0)$. This is shown also in Fig. 2.4.

2.2. LINEAR SUPERPOSITION

On readback, the existing magnetic field (that was caused by the recorded magnetization of the surface) is extremely weak, and hence, the magnetic read head will behave very nearly as a linear element. In this case, each pulse can be considered separately, and the composite readback waveform (as seen in an oscilloscope) consists of the addition of these separate, isolated pulses. This is illustrated in Fig. 2.5. where two adjacent transitions have been located at such distance from each other that the read pulses do not interfere with each other at all. That is, the maximum for each pulse occurs exactly at the same location (in time) where the transitions were written.

However, in a relentless effort to increase the magnetic recording density, the transitions are written closer and closer to each other. When this is done, two continuous pulses read, and linear superposition applied, the composite waveform looks like the one shown in Fig. 2.6. The end result (compared with Fig. 2.5.) consists of a waveform smaller than the original isolated pulse. This amplitude reduction imposes more rigorous requirements upon the sensing electronics, that is, in further amplification and noise reduction. These undesirable effects are enhanced as the density is increased.

At the same time, the maxima of the composite waveform are more separated from each other than the isolated pulses were originally. This undesirable effect is called "bit shift." A certain amount of bit shift can be compensated for by a technique called "write precompensation." This can be done by knowing "a priori" where the composite waveform would have its maxima due to simple linear superposition. By knowing this, it now becomes possible to adjust the flux reversals as they are being

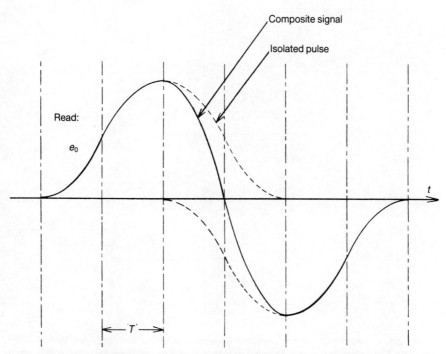

Fig. 2.5. Two adjacent bits located at the minimum distance possible between them thus totally avoiding interference.

written. Flux reversals that will provide a readback pulse with an early maximum are written late. Flux reversals that will provide a readback pulse with a late maximum are written early. This is depicted graphically in Fig. 2.7.

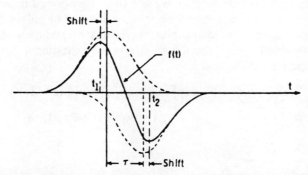

Fig. 2.6. Two adjacent bits so close together that their "tails" affect each other's amplitude and position thus causing bit shift.

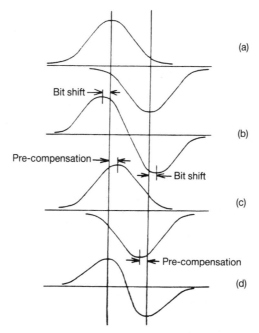

Fig. 2.7. Read waveform without and with write pre-compensation. (a) Two isolated bits. (b) Composite waveform showing bit shift. (c) Two isolated bits for which write pre-compensation took place. (d) Composite waveform showing no bit shift.

Nevertheless, let us examine these two phenomena in a more analytical way. To study bit amplitude reduction (or bit crowding) due to the superposition of pulses, let us approximate each isolated readback pulse by the Gaussian pulse $\exp(-t^2)$ as shown in Fig. 2.8. (Of course, this is not the only approximation possible, as will be mentioned in Section 2.3.) At the same time, let us consider an infinite number of pulses written at a constant spacing τ. The composite waveform $f(t)$ shown in Fig. 2.9. (which is obtained during readback) is given by the linear superposition of the individual pulses, that is:

$$f(t) = \exp(-t^2) - \exp[-(t-\tau)^2] + \exp[-(t-2\tau)^2]$$
$$- \exp[-(t+\tau)^2] + \exp[-(t+2\tau)^2] \ldots$$

$$= \sum_{n=-\infty}^{\infty} (-1)^n \exp[-(t-n\tau)^2]$$

$$n = 0, \pm 1, \pm 2, \pm 3, \ldots .$$

$$(2.2)$$

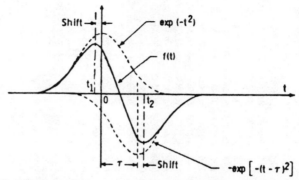

Fig. 2.8. Bit shift for two adjacent bits when the origin is placed at the maximum of one of the isolated bits. Each isolated readback pulse has been approximated by a Gaussian curve.

There is a peak at $t = 0$ with a value

$$f(0) = 2 \sum_{n=0}^{\infty} (-1)^n \exp[-(n\tau)^2] - 1.$$

Since the maximum of an isolated, normalized pulse is 1, the bit amplitude percentage is given by

$$\text{Bit amplitude} = \left\{ 2 \sum_{n=0}^{\infty} (-1)^n \exp[-(n\tau)^2] - 1 \right\} \times 100, \qquad (2.3)$$

which is the amplitude equation. This expression has been programmed

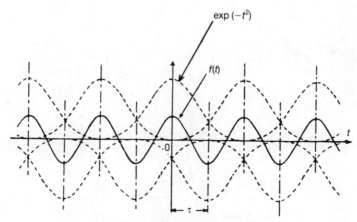

Fig. 2.9. Sequence of readback pulses.

Table 2.1

Numerical Solution of the Bit Shift and Amplitude Equations.

τ	T/τ	$\lvert t_1 \rvert$	Bit shift %	Amplitude %
4.44	1.00065	0.00000	0.00000	100.00000
3.40	1.30673	0.00003	0.00191	99.99809
3.20	1.38840	0.00011	0.00714	99.99286
3.10	1.43319	0.00021	0.01339	99.98659
3.00	1.48096	0.00037	0.02463	99.97532
2.90	1.53203	0.00064	0.04437	99.95547
2.80	1.58674	0.00110	0.07828	99.92127
2.70	1.64551	0.00183	0.13522	99.86353
2.60	1.70880	0.00297	0.22855	99.76815
2.40	1.85120	0.00732	0.61030	99.36977
2.30	1.93169	0.01107	0.96287	98.99164
2.28	1.94863	0.01199	1.05179	98.89491
2.26	1.96588	0.01297	1.14777	98.78988
2.24	1.98343	0.01401	1.25129	98.67591
2.22	2.00130	0.01513	1.36280	98.55238
2.20	2.01949	0.01631	1.48278	98.41858
2.18	2.03802	0.01757	1.61175	98.27380
2.16	2.05689	0.01890	1.75028	98.11727
2.14	2.07611	0.02032	1.89878	97.94819
2.12	2.09570	0.02181	2.05786	97.76572
2.10	2.11566	0.02339	2.22805	97.56895
2.08	2.13600	0.02506	2.41001	97.35699
2.06	2.15674	0.02682	2.60425	97.12883
2.04	2.17788	0.02868	2.81140	96.88347
2.02	2.19945	0.03062	3.03207	96.61986
2.00	2.22144	0.03267	3.26689	96.33688
1.98	2.24388	0.03481	3.51657	96.03338
1.96	2.26678	0.03706	3.78171	95.70817
1.94	2.29015	0.03941	4.06298	95.36002
1.92	2.31400	0.04187	4.36110	94.98764
1.90	2.33836	0.04443	4.67673	94.58971
1.88	2.36324	0.04710	5.01058	94.16487
1.86	2.38865	0.04988	5.36356	93.71170
1.84	2.41461	0.05277	5.73618	93.22877
1.82	2.44114	0.05578	6.12930	92.71460
1.80	2.46827	0.05889	6.54367	92.16766
1.78	2.49600	0.06212	6.98008	91.58643
1.76	2.52436	0.06547	7.43946	90.96931
1.74	2.55338	0.06893	7.92250	90.31472
1.72	2.58307	0.07250	8.43011	89.62104
1.70	2.61346	0.07619	8.96318	88.88663
1.68	2.64457	0.07999	9.52259	88.10986
1.66	2.67643	0.08391	10.10926	87.28906
1.64	2.70907	0.08794	10.72430	86.42260
1.62	2.74252	0.09209	11.36854	85.50884
1.60	2.77680	0.09634	12.04307	84.54616
1.58	2.81195	0.10072	12.74896	83.53297
1.56	2.84800	0.10520	13.48733	82.46771
1.54	2.88499	0.10980	14.25931	81.34888
1.52	2.92295	0.11450	15.06626	80.17503
1.50	2.96192	0.11932	15.90927	78.94480
1.48	3.00195	0.12424	16.78973	77.65689
1.46	3.04307	0.12928	17.70904	76.31013
1.44	3.08533	0.13441	18.66865	74.90347
1.42	3.12879	0.13966	19.67008	73.43601
1.40	3.17349	0.14501	20.71503	71.90700
1.38	3.21948	0.15046	21.80511	70.31590
1.36	3.26682	0.15601	22.94215	68.66238
1.34	3.31558	0.16166	24.12806	66.94635
1.32	3.36582	0.16741	25.36486	65.16800
1.30	3.41760	0.17326	26.65468	63.32785
1.28	3.47100	0.17920	27.99975	61.42675
1.26	3.52610	0.18524	29.40262	59.46592
1.24	3.58297	0.19137	30.86572	57.44704
1.22	3.64171	0.19759	32.39183	55.37225
1.20	3.70240	0.20390	33.98386	53.24416
1.18	3.76515	0.21031	35.64494	51.06600
1.16	3.83007	0.21679	37.37838	48.84154
1.14	3.89726	0.22337	39.18774	46.57522
1.12	3.96686	0.23003	41.07696	44.27216
1.10	4.03898	0.23678	43.05002	41.93820
1.08	4.11378	0.24360	45.11137	39.57992
1.06	4.19140	0.25051	47.26576	37.20468
1.04	4.27200	0.25750	49.51832	34.82063
1.02	4.35577	0.26456	51.87457	32.43669
1.00	4.44288	0.27170	54.34046	30.06256
0.98	4.53355	0.27892	56.92242	27.70867
0.96	4.62800	0.28621	59.62770	25.38611
0.94	4.72647	0.29358	62.46362	23.10654
0.92	4.82922	0.30102	65.43857	20.88209
0.90	4.93653	0.30853	68.56155	18.72516
0.88	5.04873	0.31611	71.84243	16.64824
0.86	5.16614	0.32376	75.29194	14.66370
0.84	5.28914	0.33147	78.92188	12.78345
0.82	5.41815	0.33926	82.74518	11.01864
0.80	5.55360	0.34711	86.77636	9.37935
0.78	5.69600	0.35502	91.03081	7.87415
0.76	5.84590	0.36300	95.52604	6.50975
0.74	6.00389	0.37104	100.28130	5.29064
0.72	6.17067	0.37915	105.31804	4.21875
0.70	6.34697	0.38731	110.66027	3.29314
0.68	6.53365	0.39554	116.3345	2.50996
0.66	6.73164	0.40382	122.3707	1.86231
0.64	6.94200	0.41217	128.8028	1.34049
0.62	7.16594	0.42057	135.6685	0.93228
0.60	7.40480	0.42903	143.0112	0.62351

into a computer. Some of the results are shown in Table 2.1. A plot of bit amplitude versus t/τ is given in Fig. 2.10.

To study the bit shift let us call, again, τ the interval at which the bits were written at a constant linear speed L in inches per second. The bit

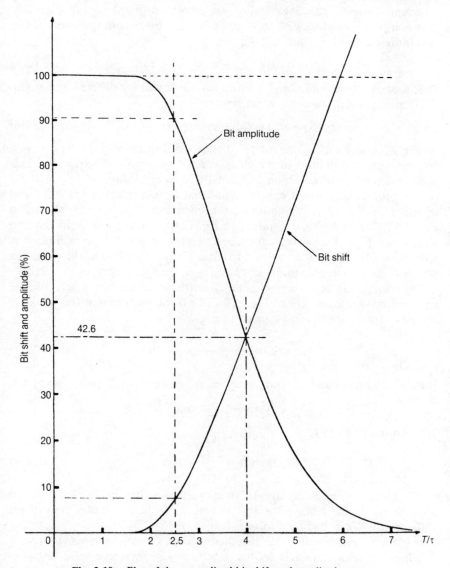

Fig. 2.10. Plot of the normalized bit shift and amplitude curves.

density is given by

$$B = \frac{1}{L\tau}.$$ (2.4)

We saw already that the worst case bit shift is obtained when two adjacent, isolated readback pulses are obtained, as shown in Fig. 2.8. The composite readback pulse is given by the linear superposition of the two individual pulses, that is

$$f(t) = \exp(-t^2) - \exp[-(t - \tau)^2].$$ (2.5)

Peaks of $f(t)$ (maximum and minimum) are found by differentiating Eq. (2.5) and equating to zero, which gives

$$t_m \exp(-t_m^2) = (t_m - \tau)\exp[-(t_m - \tau)^2],$$ (2.6)

where t_m are the abscissas of the peaks. Unfortunately, Eq. (2.6) is a transcendental equation in which t_m cannot be obtained explicitly. This equation can be solved either graphically or numerically.

The graphical solution can be visualized by considering the left-hand member of Eq. (2.6). A graph of the corresponding curve is shown in Fig. 2.11. The right-hand member of Eq. (2.6) is also a curve (as the one shown in Fig. 2.11.) but displaced to the right by an amount τ. Consequently, for a given value of τ, Eq. (2.6) is satisfied at the points in which those two curves intersect, namely, t_1 and t_2, shown in Fig. 2.12.

Therefore, the abscissas t_1 and t_2 correspond to the peaks of the composite waveform $f(t)$ of Fig. 2.8. The distance between the peaks is $|t_1| + t_2$. Total bit shift in percentage is given by

$$\text{Bit shift} = \frac{(|t_1| + t_2) - \tau}{\tau} \times 100.$$ (2.7)

But by the symmetry of our analytically approximated pulse, we have that

$$t_2 = \tau + |t_1|, \quad \text{and}$$ (2.8)

substituting (2.8) in (2.7)

$$\text{Bit shift} = \frac{2\,|t_1|}{\tau} \times 100$$ (2.9)

which is the bit shift equation. This expression has been programmed also. Some of the results are also shown in Table 2.1. and a plot of bit shift versus T/τ is included in Fig. 2.10.

It is interesting to compare the theoretical curves (denormalized) with empirical results obtained in the laboratory. The theoretical and experimental curves (for this particular experiment) are shown in Fig. 2.13.

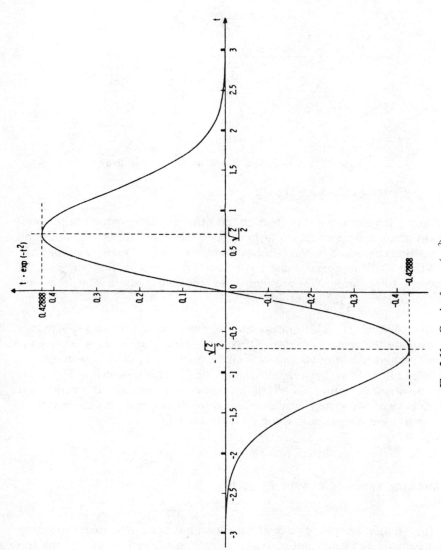

Fig. 2.11. Graph of $t \cdot \exp(-t^2)$ versus t.

33

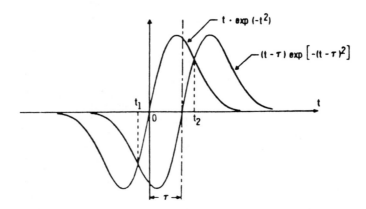

Fig. 2.12. Graphical solution of the bit shift equation.

2.3. APPROXIMATIONS

In any attempt to quantify (and qualify) the magnetic recording process, many analytical approximations have been tried both to the isolated write transition curve and to the corresponding bell shaped readback pulse.

Many such approximations have been tried with various degrees of success or failure. We are going to indicate only a few that are representative of the general trend.

(a) *Arctan* (t) This transcendental function has been used to approximate the write transition. Of course, the approximation applies for the first and fourth quadrants of a circle, namely, in the interval $-90°$ to $+90°$, or in radians, between $-\pi/2$ and $+\pi/2$. This is shown in Fig. 2.14. As mentioned before, the readback pulse is approximated by the first derivative of the write transition. Therefore, by normalizing the readback pulse to have a maximum value of 1, we have that

$$\text{Readback pulse} \propto \frac{d}{dt}[\text{Arctan}(t)] = \frac{1}{1+t^2}. \qquad (2.10)$$

Obtaining a series expansion for (2.10)

$$\text{Readback pulse} \approx 1 - t^2 + t^4 - t^6 + \ldots . \qquad (2.11)$$

(b) *Raised cosine* The cosine function has been used to approximate the isolated readback pulse. Again, this approximation applies for the first and fourth quadrants of a circle, namely, in the interval $-90°$ to $+90°$, or in radians, between $-\pi/2$ and $+\pi/2$. However, the cosine function oscillates between the values $+1$ and -1, for an amplitude of 2.

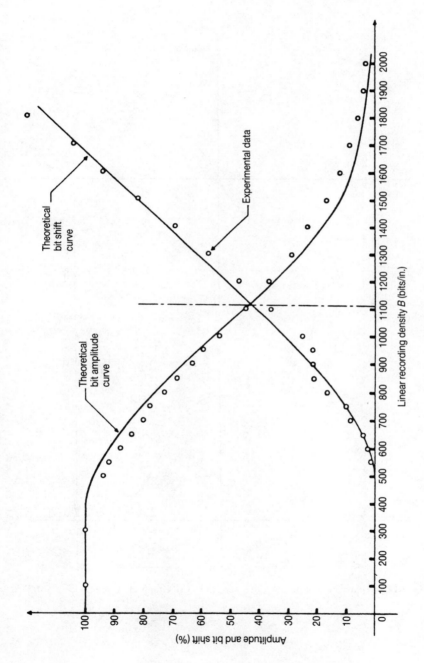

Fig. 2.13. Theoretical and experimental amplitude and bit shift curves.

35

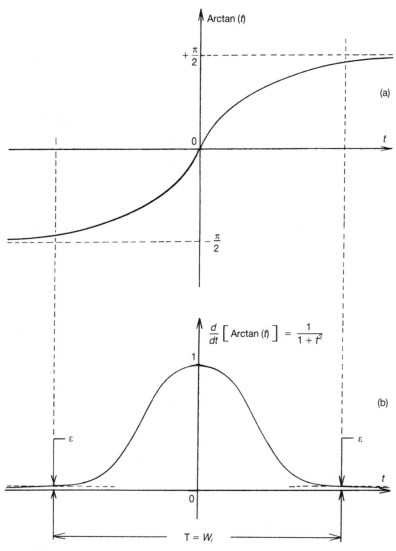

Fig. 2.14. Approximation to the write transition. (a) Graph of Arctan(*t*) vs. *t*. (b) First derivative.

Therefore, to eliminate the negative values it becomes necessary to add 1 to the function (thus the name "raised" cosine) and to normalize the approximation to a maximum value of 1. The whole expression is divided by 2, thus obtaining

$$\text{Readback pulse} \approx \tfrac{1}{2}[1 + \cos(t)]. \tag{2.12}$$

This is shown in Fig. 2.15.
Obtaining a series expansion for (2.12)

$$\text{Readback pulse} \approx 1 - \left(\frac{1}{2 \times 2!}\right)t^2 + \left(\frac{1}{2 \times 4!}\right)t^4$$

$$- \left(\frac{1}{2 \times 6!}\right)t^6 + \dots . \tag{2.13}$$

(c) *Gaussian pulse* The Gaussian pulse $\exp(-t^2)$ has been used to approximate the isolated readback pulse, that is

$$\text{Readback pulse} \approx \exp(-t^2). \tag{2.14}$$

This is shown in Fig. 2.16.
Obtaining a series expansion for (2.14)

$$\text{Readback pulse} \approx 1 - t^2 + \left(\frac{1}{2!}\right)t^4 - \left(\frac{1}{3!}\right)t^6 + \dots . \tag{2.15}$$

Notice the tremendous similarities between Eqs. (2.11), (2.13), and (2.15). Consequently, it can be said that a normalized readback pulse can be approximated by the general expression

$$\text{Readback pulse} \approx 1 - a_2 t^2 + a_4 t^4 - a_6 t^6 + \dots . \tag{2.18}$$

where the coefficients $a_2, a_4, a_6, \dots, a_{2n}$ can be determined by obtaining an empirical pulse in the laboratory and effecting a numerical approximation.

2.4. EQUALIZATION

Let us redraw Fig. 2.9. but instead of an infinite series of isolated readback pulses, let us consider a finite number of written transitions and no transitions at both ends. The result is shown in Fig. 2.17. It is seen that for the first and last pulses, due to the abscence of a pulse that could cause symmetry, the distance (in time) between the peaks of those pulses to the corresponding neighboring pulse is $T + \lambda$. That is, those pulses have

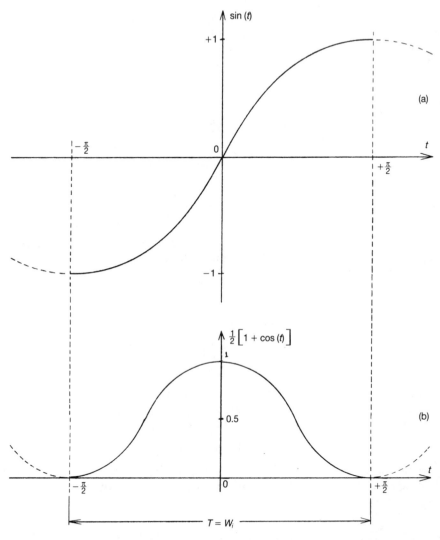

Fig. 2.15. Approximation to the read pulse. (a) Graph of sin(t) vs. t. (b) Graph of $\frac{1}{2}[1 + \cos(t)]$ vs. t.

a bit shift λ. The effect of this bit shift is an out-of-step condition with the clock signal, which in turn may cause reading errors.

It was stated before that the readback process can be considered linear. Therefore, if each individual readback pulse can be narrowed down (or slimmed) to the point where its effect upon neighboring pulses is negligible, then the bit shift λ could be eliminated. That is, the output

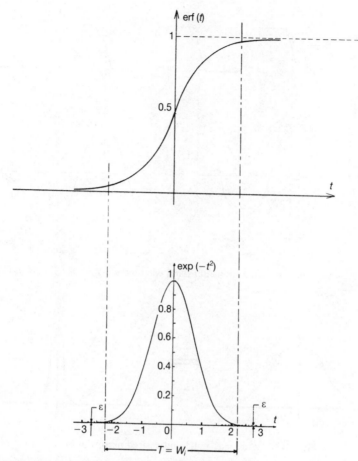

Fig. 2.16. Approximation to the read pulse. (a) Graph of erf(t) vs. t. (b) Graph of exp($-t^2$) vs. t.

pulse could be converted into a narrower pulse, as shown in Fig. 2.18. When this is done, Fig. 2.17. will be converted to Fig. 2.19. It is seen that in Fig. 2.19., due to the lack of neighboring pulse interference, the undesirable bit shift has been eliminated. As an added bonus, if this slimming down is done by a great amount (more than needed just to eliminate bit shift), the magnetic recording density could even be increased.

This pulse slimming can be done by means of a network, as shown schematically in Fig. 2.20. where the input pulse has been narrowed down by a factor K (larger than 1, of course).

This network can be specified in the frequency domain also. The

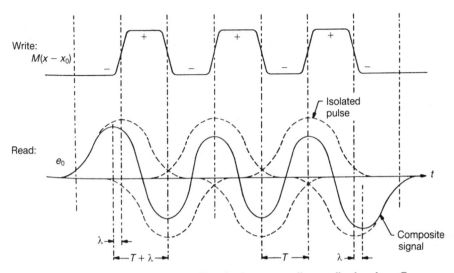

Fig. 2.17. Sequence of written transitions and corresponding readback pulses. Composite signal.

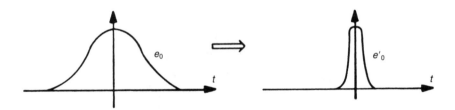

Fig. 2.18. Wide read pulse and desired narrow pulse.

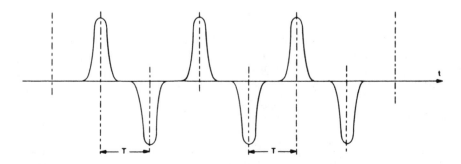

Fig. 2.19. Sequence of readback signals with narrowed pulses.

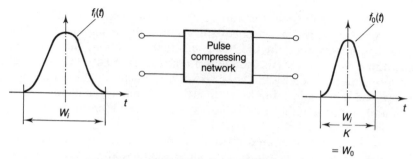

Fig. 2.20. Pulse compressing (or slimming) filter.

network (or filter) that results is called an "equalizer." A long time ago, a method was used to increase the recording density on a magnetic surface, by means of two R–C sections between the terminals of the readback head (Gabor, 1959).

The signal received from the magnetic head was thus slightly compressed (slimmed down) before entering the read amplifier. Other implementations included active elements and a tapped delay line with three operational amplifiers. Yet another method consisted of a delay line, the second derivative of the pulse, and a subtraction. As a matter of fact, the subject of equalization has received (and still is receiving) a great deal of attention throughout the years. Surely, there will be many other proposals in the future.

In the frequency domain, it is possible to say that the narrower pulse out of the network contains largely increased high frequencies compared with the wider input pulse. This implies a certain "boost" of the high frequencies, leading some people to believe that this process of attenuating some frequencies and boosting others implied an "optimum" frequency spectrum for the readback signal (thus the term "equalizer," although in reality, there is nothing to "equalize"). In reality, this is quite an optimistic exaggeration because quite different frequency spectra (not a unique one) are successful in narrowing the readback pulses. This indicates that the exact shape of the spectrum is not critical, and empirical methods are well justified for designing equalizers.

The equalizing concept can be justified in a somewhat more rigorous way: It was stated before that the readback signal can be expressed as:

$$e_0 = CV \frac{\partial}{\partial x} \int_{-\infty}^{\infty} D(x) \, M(x - x_0) \, dx, \qquad (2.17)$$

where $D(x)$ represents the sensitivity function of the R/W head. The integration is a convolution. However, we know that a filter performs the convolution (in the time domain) of its impulse response with the input pulse. Consequently, if we postulate a filter whose impulse response is equal to the reciprocal of the head sensitivity function, the convolution performed by the filter (the equalizer) tends to eliminate the convolution (or spreading) caused by the head sensitivity function. Therefore, we can summarize the whole thing by stating that the equalizer is nothing more than a filter whose impulse response is equal to the reciprocal of the head sensitivity function.

However, in the final analysis the equalizer is, at best, an approximation that cannot create things that were eliminated by the R/W head, such as high frequencies. That is, the equalizer cannot create something from nothing. In practical terms, equalization is a rather limited process circumscribed exclusively to what is already present in the readback signal. Therefore, this reshaping of the readback signal spectrum is usually done at the expense of boosting high frequency noise, thus lowering the signal-to-noise ratio. A very long time ago, the readback signal had a signal-to-noise ratio of approximately 40 dB. After 30 years of development and in spite of all the efforts spent, this has become 10 dB. We can say that we have lost 30 dB in the signal-to-noise ratio, in 30 years, or an average of 1 dB per year. If this trend continues unabated, in 10 more years, we can expect the readback signal to be "in the mud," in which case, the read channel will consist of a process very different to what we have today (see Chapter 8).

In summary, whether to "equalize" or not becomes an engineering compromise: Should we give away signal-to-noise ratio to obtain resolution? If the system has already a very low signal-to-noise ratio and we cannot afford to give away any more, then equalization is a very poor choice indeed. This decision must be made for every particular application.

We stated before that a great number of mathematical approximations have been made to the readback signal. The mathematical approximation can be used to design the equalizing network. After the filter is designed, it becomes necessary to adjust the components in the laboratory. Consequently, the whole process can be considered rather empirical, with the approximation first, and the final adjustment later. However, in between, it becomes possible to obtain a transfer function not too far from the "ball park."

The Gaussian approximation has proven as good as any other, so we will use it here. For purposes of synthesizing the pulse compression network (pulse slimming filter), let us call:

Input Pulse: $f_i(t) = \exp(-t^2)$,   and   (2.18)

Output Pulse: $f_o(t) = \exp(-K^2 t^2)$,   (2.19)

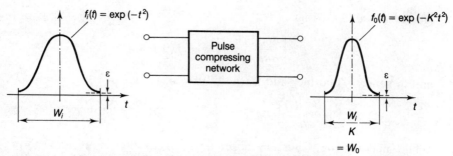

Fig. 2.21. Pulse compressing network for which the input and output signals have been approximated by Gaussian pulses.

where K is the compression factor at the base of the pulse. That is, if the input pulse has a width at the base given by W_i, then the output pulse will have a width at the base of:

$$W_o = \frac{W_i}{K} \tag{2.20}$$

where

$$K \geqslant 1. \tag{2.21}$$

For this case, Fig. 2.20. becomes transformed into Fig. 2.21. The transfer function of the filter is given by

$$H(s) = \frac{F_o(s)}{F_i(s)} \tag{2.22}$$

where

$$F_o(s) = \mathscr{L}[f_o(t)], \tag{2.23}$$

$$F_i(s) = \mathscr{L}[f_i(t)] \tag{2.24}$$

and $\mathscr{L}$ is the Laplacian operator. By using the Gaussian approximation proposed before, we have that

$$H(s) = \frac{F_o(s)}{F_i(s)} = \frac{1}{K} \exp(-\varphi^2 s^2) \tag{2.25}$$

where

$$\varphi = \frac{\sqrt{K^2 - 1}}{2K}. \tag{2.26}$$

Let us make a series expansion of (2.25)

$$\frac{F_o(s)}{F_i(s)} = b_o - b_2 s^2 + b_4 s^4 - b_6 s^6 + \ldots$$

or

$$F_o(s) = b_o F_i(s) - b_2 s^2 F_i(s) + b_4 s^4 F_i(s) - \ldots. \tag{2.27}$$

However, we know that

$$\mathcal{L}\left[\frac{df(t)}{dt}\right] = sF(s) - f(0+)$$

and in general (Gardner and Barnes, 1956, p. 129),

$$\mathcal{L}\left[\frac{df^{(n)}(t)}{dt}\right] = s^n F(s) - \sum_{k=1}^{n} f^{(k-1)}(0+)s^{n-k}. \qquad (2.28)$$

While making the Gaussian approximation to the input and output pulses, for convenience, they were centered at the origin. Nevertheless, this is not indispensable, and a similar transfer function would have been obtained if both pulses had been centered very far from the origin. Consequently, it is possible to say that

$$f_i^I(0+) = f_i^{II}(0+) = f_i^{III}(0+) = \ldots = 0.$$

Obtaining the inverse Laplace transform of both members in Eq. (2.27),

$$f_o(t) = b_o f_i(t) - b_2 f_i^{II}(t) + b_4 f_i^{IV}(t) - b_6 f_i^{VI}(t) + \ldots . \qquad (2.29)$$

This is shown in Fig. 2.22.

It is not surprising at all that some approaches to the pulse slimming filter synthesis in the time domain have used delay lines, obtained the second derivative of the incoming pulse $f_i(t)$, and then performed a subtraction. Obviously, they were truncating the right-hand member of Eq. (2.29) after only two terms, thus effecting quite an approximation.

However, that is not the only approximation possible. The magnitude functions for the input and output pulses, plus the impulse response of the desired network, are shown in Fig. 2.23. It is seen that in the frequency domain, any rational approximation to the transcendental (2.25) will hold only up to a frequency ω_d. Thereafter, the approximation will be very far from the desired frequency response. However, if the approximation possesses linear phase up to ω_d, the resulting pulse might not be too distorted, in which case the rational approximation could be used.

Equation (2.25) has been approximated by means of a continuous fraction expansion, because if offers very fast convergence. A table of the approximations for $H(s)H(-s)$ is given as Table 2.2, together with the corresponding poles and zeros. The approximations given in that table are the diagonal entries (or staircase entries) of the Padé table for $H(s)H(-s)$. It is now possible to take from each entry the left-hand plane poles only (thus obtaining a Hurwitz polynomial in the denominator) and the right-hand plane zeros only (for phase linearity) and

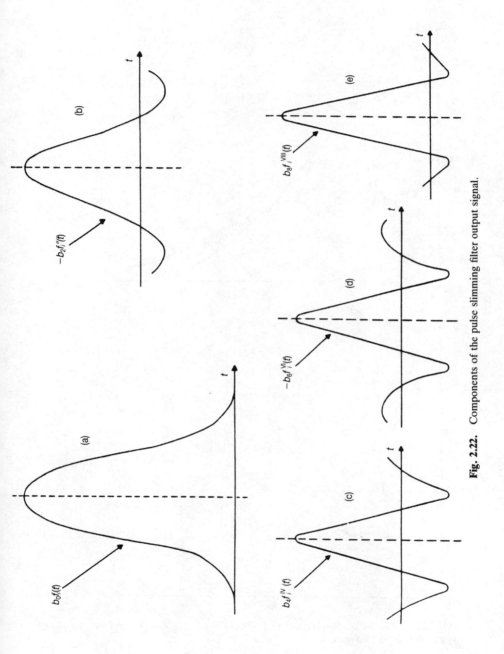

Fig. 2.22. Components of the pulse slimming filter output signal.

45

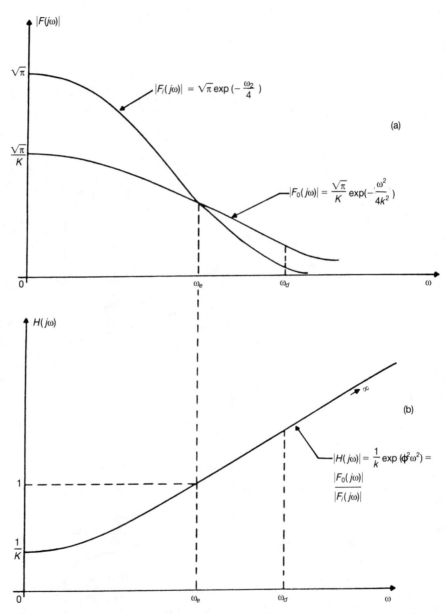

Fig. 2.23. Curves of the magnitude function. (a) For the input and output pulses. (b) For the desired network.

Table 2.2.

Approximations to $H(s)H(-s)$.

N	$H(s)H(-s)$	Zeros	Poles
2	$\left(\dfrac{1}{K^2}\right)\dfrac{1}{1+2\phi^2 s^2}$		$\dfrac{1}{\phi}(0\pm j0.7071\,0678)$
3	$\left(\dfrac{1}{K^2}\right)\dfrac{1-\phi^2 s^2}{1+\phi^2 s^2}$	$\dfrac{1}{\phi}(\pm1+j0)$	$\dfrac{1}{\phi}(0\pm j1)$
4	$\left(\dfrac{1}{K^2}\right)\dfrac{3-2\phi^2 s^2}{3+4\phi^2 s^2+2\phi^4 s^4}$	$\dfrac{1}{\phi}(\pm1.22474487+j0)$	$\dfrac{1}{\phi}(\pm0.3352200067\pm j1.054690705)$
5	$\left(\dfrac{1}{K^2}\right)\dfrac{3-3\phi^2 s^2+\phi^4 s^4}{3+3\phi^2 s^2-\phi^4 s^4}$	$\dfrac{1}{\phi}(\pm1.271229882\pm j0.340625032)$	$\dfrac{1}{\phi}(\pm0.340625032\pm j1.271229882)$
6	$\left(\dfrac{1}{K^2}\right)\dfrac{15-12\phi^2 s^2+3\phi^4 s^4}{15+18\phi^2 s^2+9\phi^4 s^4+2\phi^6 s^6}$	$\dfrac{1}{\phi}(\pm1.45534667\pm j0.343560805)$	$\dfrac{1}{\phi}(0\pm j1.34867233)$ $\dfrac{1}{\phi}(\pm0.587391538\pm j1.29829514)$
7	$\left(\dfrac{1}{K^2}\right)\dfrac{15-15\phi^2 s^2+6\phi^4 s^4-\phi^6 s^6}{15+15\phi^2 s^2+6\phi^4 s^4+\phi^6 s^6}$	$\dfrac{1}{\phi}(\pm1.52387182+j0)$ $\dfrac{1}{\phi}(\pm1.47994076\pm j0.5927200616)$	$\dfrac{1}{\phi}(0\pm j1.52387182)$ $\dfrac{1}{\phi}(\pm0.5927200616\pm j1.47994076)$
8	$\left(\dfrac{1}{K^2}\right)\dfrac{105-90\phi^2 s^2+30\phi^4 s^4-4\phi^6 s^6}{105+120\phi^2 s^2+60\phi^4 s^4+16\phi^6 s^6+2\phi^8 s^8}$	$\dfrac{1}{\phi}(\pm1.68054 8372+j0)$ $\dfrac{1}{\phi}(\pm1.64112335\pm j0.596160249)$	$\dfrac{1}{\phi}(\pm0.79700116\pm j1.497202089)$ $\dfrac{1}{\phi}(\pm0.250044024\pm j1.56720106)$
9	$\left(\dfrac{1}{K^2}\right)\dfrac{105-105\phi^2 s^2+45\phi^4 s^4-10\phi^6 s^6+\phi^8 s^8}{105+105\phi^2 s^2+45\phi^4 s^4+10\phi^6 s^6+\phi^8 s^8}$	$\dfrac{1}{\phi}(\pm1.6572801+j0.801741003)$ $\dfrac{1}{\phi}(\pm1.72038868\pm j0.252045949)$	$\dfrac{1}{\phi}(\pm0.801741003\pm j1.6572801)$ $\dfrac{1}{\phi}(\pm0.252045949\pm j1.72038868)$

construct a rational approximation (ratio of polynomials) to $H(s)$, thus making sure that the resulting filter is realizable.

However, while doing the filter denormalization, it becomes necessary to know one of the denormalizing factors, namely, the width at the base of our approximation. Unfortunately, the Gaussian pulse is asymptotic, that is, the width spans a range from $-\infty$ to $+\infty$. But the pulses coming from the read head have a finite width at the base, albeit difficult for them to be measured. Some applications can use the bit amplitude at the 50% level. However, in our case this is not possible because the pulse slimming filters have been predicated as reliefs for bit shift, this undesirable effect is caused by the "tails" of the pulses, and they are at the base, not at the rather harmless 50% level.

For purposes of finding W_i let us go back to Fig. 2.21. and consider the same input and output pulses. The magnitude of the Fourier transform for the input and output pulses to the network are given by

$$|F_i(j\omega)| = \sqrt{\pi}\exp\left(-\frac{\omega^2}{4}\right), \quad \text{and} \tag{2.30}$$

$$|F_o(j\omega)| = \frac{\sqrt{\pi}}{K}\exp\left(-\frac{\omega^2}{4K^2}\right). \tag{2.31}$$

A graph of Eqs. (2.30) and (2.31) is given in Fig. 2.24. Let us consider the frequency ω_e at which the magnitudes of the input and output are equal. This is obtained by equating expressions (2.30) and (2.31)

$$\sqrt{\pi}\exp\left(-\frac{\omega_e^2}{4}\right) = \frac{\sqrt{\pi}}{K}\exp\left(-\frac{\omega_e^2}{4K^2}\right),$$

which gives

$$\omega_e = 2K\sqrt{\frac{\ln K}{K^2 - 1}}. \tag{2.32}$$

A graph of Eq. (2.32) is given in Fig. 2.25.

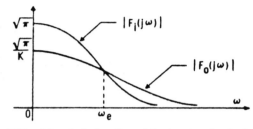

Fig. 2.24. Magnitude function of the input and output pulses.

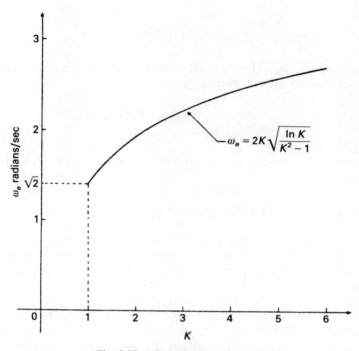

Fig. 2.25. Graph of ω_e versus K.

To eliminate the radical of (2.32), let us square both members:

$$\omega_e^2 = 4K^2 \frac{\ln K}{K^2 - 1}. \tag{2.33}$$

Let us now consider the case where the network does not compress the pulse at all, that is, let us make $K = 1$. In other words, we want to determine the frequency ω_e when we have an output pulse with the same width as the input pulse. In this limiting case, the network would act as an all-pass network which would have at its input an asymptotic pulse and which would reproduce the pulse faithfully, but with a frequency limitation such that the output pulse would have a finite width.

For $K = 1$, Eq. (2.33) is indeterminate. But applying l'Hôpital's rule, we obtain

$$\lim_{K \to 1} \omega_e^2 = \lim_{K \to 1} \left[\frac{4K^2 \ln K}{K^2 - 1} \right] = 2. \tag{2.34}$$

Then

$$\lim_{K \to 1} \omega_e^2 = 2.$$

That is,

$$\lim_{K \to 1} \omega_e = \sqrt{2} \text{ radians/sec,}$$

as shown in Fig. 2.25. Therefore, the output width of the pulse is nothing more than the period of this frequency; that is,

$$W_i = \frac{2\pi}{\lim\limits_{K \to 1} \omega_e} = \frac{2\pi}{\sqrt{2}} = \sqrt{2}\,\pi = 4.4428829 \text{ sec.} \qquad (2.35)$$

This denormalizing factor will be used in our subsequent implementations (see Chapter 8).

2.5. FOURIER COMPONENTS

It is possible to write a string of ones and zeros on a magnetic surface, then, during readback, locate a spectrum analyzer at the output of the read head, thus obtaining, empirically, the frequency components that the read channel will have to handle. However, sometimes it would be useful to know beforehand what those frequencies will be. Later on, it will be possible to corroborate those findings in the laboratory, as indicated above. Instead of dealing in generalities, let us concentrate on one particular example and establish the methodology. Thereafter, it should be possible to apply the method to any particular application.

Example. Let us consider a magnetic recording system having a bit window 4 microseconds wide. Figure 2.26. shows a magnetic recording system in which the ones and zeros have been written in an alternate fashion. To the naked eye the composite readback waveform looks like a pure sinusoid (although it is composed of the linear superposition of isolated pulses) with a period of 4×4 microseconds $= 16$ microseconds, that is, with a frequency

$$\frac{1}{16 \times 10^{-6}} = 62.5 \text{ KHz.} \qquad (2.36)$$

This frequency has been dubbed $1F$. In reality, the readback waveform, although looking like a perfect sinusoid, is not. This can be verified very easily by placing a spectrum analyzer at the output of the read head and noticing that there are a tremendous number of harmonics. Unfortunately, the name (although not quite accurate) has persisted and is used all the time. In reality, it is better to say that $1F$ is the fundamental (or first harmonic) of all the Fourier components of the read waveform for the

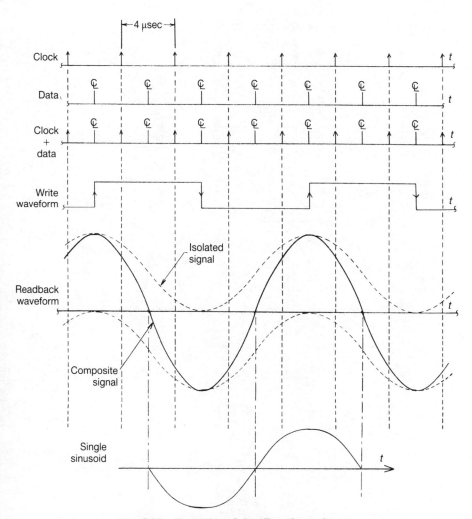

Fig. 2.26. Formation of the 1F read waveform.

case of alternate ones and zeros. Notice the similarities between Figs. 2.26. and 2.5.: In both figures, the write transitions have been placed at such distance that the isolated readback pulses do not interfere with each other at all (neither bit shift nor crowding).

It is possible to place the write transitions much closer together in which case the readback waveform will look like the one shown in Fig. 2.27. For this case, the fundamental (or first harmonic) of all the Fourier

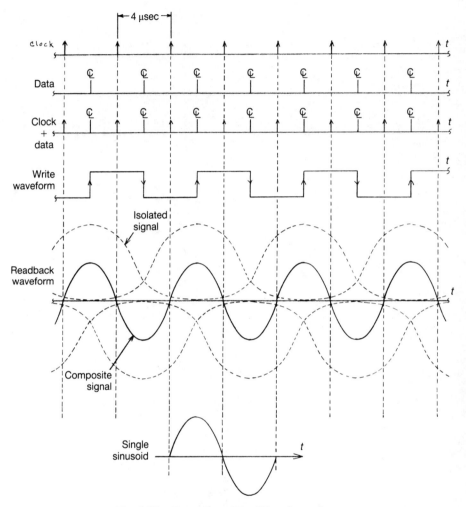

Fig. 2.27. Formation of the 2F read waveform.

components consists of a frequency

$$2F = \frac{1}{8 \times 10^{-6}} = 125 \text{ KHz}. \tag{2.37}$$

In other cases, the write transitions are not as close as in Fig. 2.27., nor as far as in Fig. 2.26. For this intermediate case, the first harmonic has a

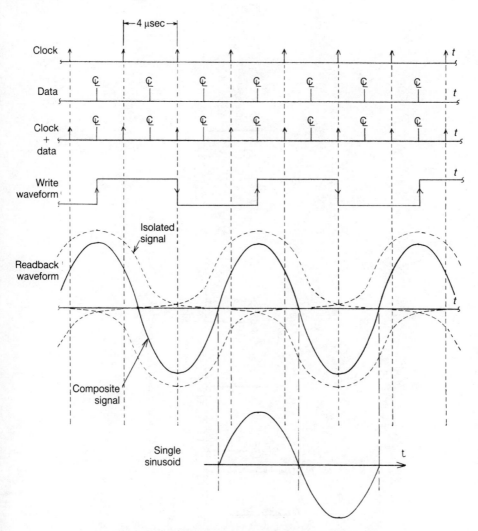

Fig. 2.28. Formation of the (4/3)F read waveform.

frequency

$$\left(\frac{4}{3}\right)F = \frac{1}{12 \times 10^{-6}} = 83.333333 \text{ KHz},\qquad(2.38)$$

which is shown in Fig. 2.28.

The method is quite general. Therefore, preparing a set of graphs as

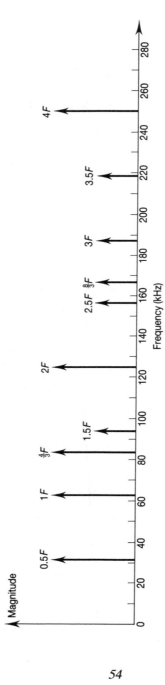

Fig. 2.29. An example of a readback frequency spectrum.

those shown in Fig. 2.26., 2.27. and 2.28., but for all the possible patterns of a given code (Chapters 6 and 7), results in a set of frequencies located in a theoretical frequency spectrum such as the one shown (for example purposes only) in Fig. 2.29. In this spectrum, and arbitrarily, we have shown all the fundamental frequencies with one magnitude and the harmonics with another. This is similar (except in magnitude) to the readings taken in the laboratory with a spectrum analyzer.

However, a theoretical spectrum such as that shown in Fig. 2.29. serves two purposes:

(a) To establish beforehand the frequency range of the read/write channel (Chapter 8), and,
(b) Comparing this theoretical and clean! graph with the output of a spectrum analyzer to help determine how noisy (from a frequency point of view) the system is, up to that point.

2.6. VERTICAL MAGNETIC RECORDING

In today's digital magnetic recording systems, the particles get deposited on the disks in a horizontal fashion, that is, parallel to the disk substrates. The R/W heads associated with this particle orientation are called "ring heads." For a long time, it was found that orienting the magnetic particles with their axis of anisotropy perpendicular to the substrate would result in "standing room only" particles, thus providing much larger recording densities than heretofore possible. The first efforts along the so-called vertical magnetic recording concept involved the development of "tunnel heads" (as opposed to ring heads), and they were not very successful. However, this concept has been receiving a great deal of interest again and some very astonishing results have been reported, even by using common ring heads. Linear densities as high as 300,000 bits/inch have been obtained. Therefore, the extrapolation (and corresponding prediction) shown in Fig. 2.1. seem attainable. It is very possible that vertical magnetic recording will play an extremely important role in the development of future disk drives.

It is possible to decompose a magnetic particle into two components, horizontal and vertical. Thereafter, by considering each component as a vector, the original particle can be perceived as the resultant or vector addition of those two components. This is shown graphically in Fig. 2.30a. Consequently, according to the value of the angle θ (the axis of

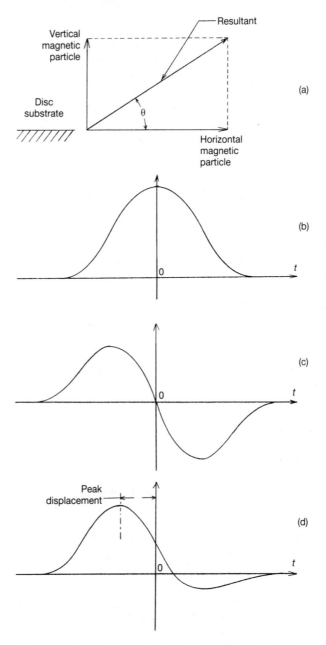

Fig. 2.30. Intermediate recording. (a) Vector addition. (b) Purely horizontal readback pulse. (c) Purely vertical readback pulse. (d) Addition of the two (2) previous pulses.

anisotropy), it is possible to make the following classification

$\theta = 0°$:   horizontal magnetic recording,

$0° < \theta < 45°$: quasi-horizontal magnetic recording,

$\theta = 45°$: hybrid magnetic recording,

$45° < \theta < 90°$: quasi-vertical magnetic recording,   and

$\theta = 90$: vertical magnetic recording.

We considered already the readback pulse obtained from a purely horizontal magnetic recording system, and it is shown in Fig. 2.30b. The readback pulse obtained from a purely vertical recording system is shown in Fig. 2.30c. The relative magnitude of these two pulses depend on the angle θ. Therefore, by adding these two pulses, we obtain a signal such as that shown in Fig. 2.30d, which is typical of the not-so-pure vertical recording pulses obtained with most systems today.

Obviously the amount of pulse asymmetry (or distortion) has a good correlation with the value of θ.

Equalizer.   The pulse obtained from a purely vertical magnetic recording system, as shown in Fig. 2.30c. can be approximated by the first derivative of the horizontal recording readback pulse. That is, by calling

$f(t)$ = readback pulse obtained from a purely
horizontal magnetic recording system   and

$g(t)$ = readback pulse obtained from a purely
vertical magnetic recording system,

then

$$g(t) \cong \frac{df(t)}{dt}.$$

(2.39)

Let us consider a pulse slimming filter (equalizer) to be used in a vertical magnetic recording system. Let us call

$g_i(t)$ = input pulse to the slimming filter   and

$g_o(t)$ = output pulse from the slimming filter.

Applying Eq. (2.39),

$$g_i(t) = \frac{df_i(t)}{dt}, \quad \text{and}$$

(2.40)

$$g_o(t) = \frac{df_o(t)}{dt}.$$

(2.41)

The transfer function of the vertical magnetic recording filter is given by

$$G(s) = \frac{\mathscr{L}[g_o(t)]}{\mathscr{L}[g_i(t)]} \,. \tag{2.42}$$

Substituting (2.40) and (2.41) in (2.42)

$$G(s) = \frac{\mathscr{L}\left[\dfrac{df_o(t)}{dt}\right]}{\mathscr{L}\left[\dfrac{df_i(t)}{dt}\right]} \,. \tag{2.43}$$

Applying to (2.43) the Laplace transform for the first derivative of a function (Gardner and Barnes, 1956, p. 128),

$$G(s) = \frac{sF_o(s) - f_o(0+)}{sF_i(s) - f_i(0+)} \,. \tag{2.44}$$

Again, by considering that the input and output pulses are far from the origin, we can say

$$f_o(0+) = f_i(0+) = 0. \tag{2.45}$$

Substituting (2.45) in (2.44)

$$G(s) = \frac{sF_o(s)}{sF_i(s)} = \frac{F_o(s)}{F_i(s)} \,. \tag{2.46}$$

Comparing Eqs. (2.46) and (2.22)

$$G(s) = H(s) = \frac{F_o(s)}{F_i(s)} \,, \tag{2.47}$$

which means that the transfer functions for the equalizers used in horizontal and vertical magnetic recording are exactly alike.

2.7. SUMMARY

This has been an introduction to the extensive subject of magnetic recording. For the sake of brevity, digital recording alone (not analog recording) has been considered. Even then, just a few topics have been considered that directly affect the performance of disk drives. However, the subjects discussed above are of vital importance, not only to the DASD designer, but also to the user, who must become aware of the strengths and limitations imposed by this technology. The limitations of magnetic recording constitute the motivating force behind the continuing effort to investigate alternative methods of recording. Presently, these efforts are starting to bear fruit as products that incorporate other methods of recording, such as optical recording, continue to be announced. Those interested in a deeper understanding of optical and magnetic recording are encouraged to consult the Bibliography.

3

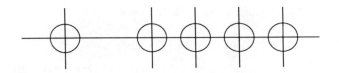

Formulas
and
Constraints

There are a number of mathematical equations that serve not only to guide the construction of DASDs but also to evaluate their performance. By becoming aware of these formulas, computer scientists can better reconfigure the data in DASDs intended to be incorporated into a data processing environment. This chapter offers a collection of such formulas. For purposes of illustration, a specific example, that uses all the formulas, has been included. Since the formulas are so numerous, for purposes of simplification some of them have been included in a graph that illustrates, at a glance, the interplay of several important parameters.

3.0. ARCHITECTURE

The word "architecture" is usually associated with only one of its interpretations, namely, the art and practice of designing and building structures, specifically, habitable ones. However, when dealing with disk drives, architecture refers to a formation or construction which is the result of a conscious act. In this regard, we are not going to deal with the details of packaging and/or implementation, but instead with the acts of calculating, defining, and planning.

3.1. FORMULAS

The following formulas are given without derivations for the following reasons:

(A) Some are mere definitions.
(B) Others are straightforward.
(C) The remaining ones are easily derived.

They are presented here, mostly, for purposes of completeness.

(1) Number of bits per byte

$$\text{byte} = 8 \text{ bits}. \tag{3.1}$$

(2) Period of one disk revolution

$$t_r = \frac{1}{S}. \tag{3.2}$$

(3) Latency, or period of one-half disk revolution

$$Y = \frac{t_r}{2} = \frac{1}{2S}. \tag{3.3}$$

(4) For Run Length Limited Codes RLL (d, k)

(a)

$$DR = \frac{\text{number of bits per inch}}{\text{number of flux changes per inch}} = \frac{BPI}{FCI} = \frac{B}{F}$$

$$= (d + 1)\left(\frac{m}{n}\right), \tag{3.4}$$

(b)

$$FR = \frac{\text{highest frequency}}{\text{lowest frequency}} = \frac{F_H}{F_L} = \frac{k + 1}{d + 1}, \tag{3.5}$$

(c)

$$w = \left(\frac{m}{n}\right) \times 100. \tag{3.6}$$

(5) Relationship between the data rate Δ_B in the controller interface to the data rate Δ_F at the R/W head

$$\Delta_B = (DR)(\Delta_F). \tag{3.7}$$

(6) Bit density

$$B = BPI = (DR)(FCI) = (DR)(F). \tag{3.8}$$

(7) Maximum areal density in bits per square inch

$$A_{max} = BT = (DR)(F)(T). \tag{3.9}$$

(8) Number of bits per track at the interface to the controller

$$N_B = B(\pi D_I). \tag{3.10}$$

(9) Number of bytes per track at the interface to the controller

$$N_{BY} = \left[\frac{1}{8}\right] B(\pi D_I). \tag{3.11}$$

(10) Rotational speed

$$S = \frac{\Delta_B}{B(\pi D_I)}. \tag{3.12}$$

(11) Substituting (3.10) into (3.11) and rearranging terms,

$$\Delta_B = S \cdot N_B, \tag{3.13}$$

Which is a very important relationship between three track constants. That is, only two of them can be independent, not the three of them.

(12) Total number of flux transitions per track at the R/W head

$$N_F = \frac{N_B}{(DR)} = \frac{\Delta_B}{S(DR)}. \tag{3.14}$$

(13) Linear length of innermost track

$$G_I = \pi \cdot D_I. \tag{3.15}$$

(14) Linear length of outermost track

$$G_O = \pi \cdot D_O. \tag{3.16}$$

(15) Radial width of data band (distance travelled by the R/W head during maximum seek)

$$H = \frac{D_O - D_I}{2}. \tag{3.17}$$

(16) Track pitch

$$P_T = \frac{1}{T}. \tag{3.18}$$

(17) Total number of tracks per disk side
$$K = T \cdot H. \tag{3.19}$$

(18) "Alternate ones" fundamental frequency at the R/W head
$$1F = \frac{\Delta_F}{4}. \tag{3.20}$$

(19) "All ones" fundamental frequency at the R/W head
$$2F = \frac{\Delta_F}{2}. \tag{3.21}$$

(20) Data cell at the interface to the controller
$$T_C = W_{FB} = \frac{1}{\Delta_B}. \tag{3.22}$$

(21) Half-period of the lowest frequency at the R/W head
$$T_{FL} = (k + 1) \times \frac{m}{n} \times T_C. \tag{3.23}$$

(22) Half-period of the highest frequency at the R/W head
$$T_{FH} = (d + 1) \times \frac{m}{n} \times T_C. \tag{3.24}$$

(23) Lowest frequency at the R/W head
$$F_L = \frac{1}{2T_{FL}}. \tag{3.25}$$

(24) Highest frequency at the R/W head
$$F_H = \frac{1}{2T_{FH}}. \tag{3.26}$$

(25) Minimum spacing between flux changes
$$W_{F_{(minimum)}} = T_{FH}. \tag{3.27}$$

(26) Maximum spacing between flux changes
$$W_{F_{(maximum)}} = T_{FL}. \tag{3.28}$$

(27) Half-space between bits at the controller interface
$$T_W = \frac{T_C}{2} = \frac{W_{FB}}{2} = \frac{1}{2\Delta_B}. \tag{3.29}$$

(28) Unformatted capacity per disk side in bits

$$C_B = K \cdot N_B. \tag{3.30}$$

(29) Unformatted capacity per disk side in bytes

$$C_{BY} = \frac{C_B}{8}. \tag{3.31}$$

(30) Unformatted total capacity per drive in bits

$$C_T = (DS)C_B. \tag{3.32}$$

(31) Unformatted total capacity per drive in bytes

$$C_{TY} = \frac{C_T}{8}. \tag{3.33}$$

3.2. EXAMPLE

Let us familiarize ourselves with the previous formulas. For this purpose, let us consider the following parameters:

$$D_I = 2 \text{ inches}$$

$$D_O = 3.74 \text{ inches}$$

$$\text{Recording code RLL } (2, 7)$$

$$\Delta_B = 24 \text{ million bits/sec}$$

$$S = 6,000 \text{ RPM} = 100 \text{ RPS}$$

$$T = 1,700 \text{ tracks/inch}$$

This point has been plotted in Fig. 3.1. and we can deduce that such a product could have an FCS of 1Q91.

(1) Period of one disk revolution

$$t_r = \frac{1}{S} = \frac{1}{100} = 10 \text{ milliseconds.} \tag{3.34}$$

(2) Latency

$$Y = \frac{1}{2S} = \frac{1}{2 \times 100} = 5 \text{ milliseconds.} \tag{3.35}$$

(3) For the RLL $(2, 7)$ code and the chosen encode/decode table:

$$d = 2, \quad k = 7, \quad m = 2, \quad n = 4, \quad r = 3;$$

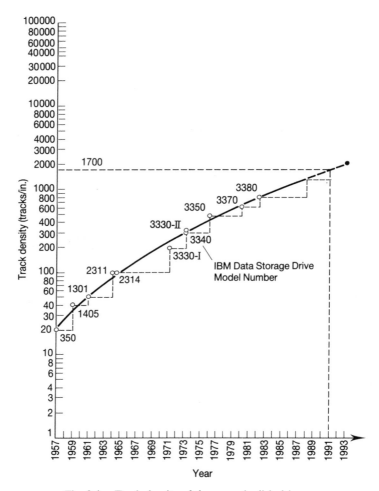

Fig. 3.1.   Track density of the example disk drive.

(a)

$$DR = (d + 1)\frac{m}{n} = (2 + 1)\tfrac{2}{4} = 1.5; \qquad (3.36)$$

(b)

$$FR = \frac{k + 1}{d + 1} = \frac{7 + 1}{2 + 1} = \tfrac{8}{3} = 2.67; \qquad (3.37)$$

(c)

$$w = \frac{m}{n} \times 100 = \tfrac{2}{4} \times 100 = 50\%; \qquad (3.38)$$

(4) Data rate at the R/W head

$$\Delta_F = \frac{\Delta_B}{(DR)} = \frac{24 \times 10^6}{1.5} = 16 \text{ megabits/sec.} \qquad (3.39)$$

(5) Number of bits per track at the interface to the controller

$$N_B = \frac{\Delta_B}{S} = \frac{24 \times 10^6}{100} = 240,000 \text{ bits/track.} \qquad (3.40)$$

(6) Number of bits per inch at the innermost track

$$B = \frac{\Delta_B}{(\pi D_1)S} = \frac{24 \times 10^6}{\pi \times 2 \times 100} = 38,197 \text{ bits/inch.} \qquad (3.41)$$

This point has been plotted in Fig. 3.2., and we can deduce that such a product could have an FCS of 1Q91.

(7) Maximum areal density

$$A_{max} = BT = 38,197 \times 1,700 = 64,934,800 \text{ bits/square inch.} \qquad (3.42)$$

This point has been plotted in Fig. 3.3., and we can deduce that such a product could have an FCS of 1Q91.

(8) Ratio between bits/inch to tracks/inch

$$\frac{B}{T} = \frac{38,197}{1,700} = 22.45 \qquad (3.43)$$

This point has been plotted in Fig. 3.4., and we can deduce that such a product could have an FCS of 3Q89.

(9) Flux density at the innermost track

$$F = \frac{B}{(DR)} = \frac{38,197}{1.5} = 25,465 \text{ flux changes/inch.} \qquad (3.44)$$

(10) Let us assume a product with an FCS of 1Q91. Entering this date in Fig. 3.5., we can deduce that such a product should have an average seek time of 13 millisceonds.

(11) Number of bytes per track at the interface to the controller

$$N_{BY} = \frac{N_B}{8} = \frac{240,000}{8} = 30,000 \text{ bytes/track.} \qquad (3.45)$$

(12) Total number of flux transitions per track at the R/W head

$$N_F = \frac{N_B}{(DR)} = \frac{240,000}{1.5} = 160,000 \text{ flux changes/track.} \qquad (3.46)$$

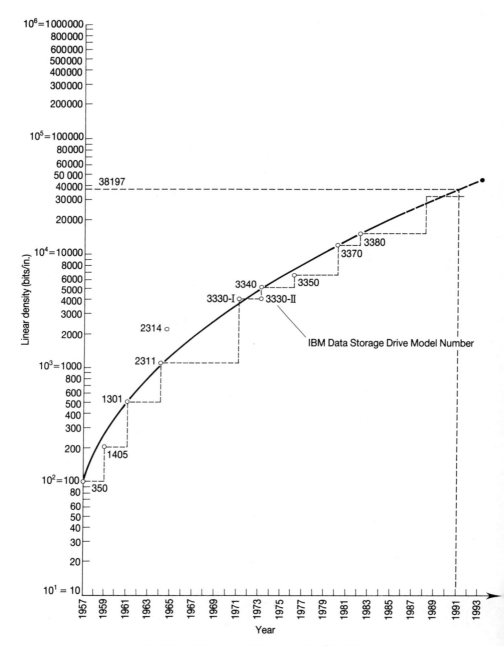

Fig. 3.2.   Bit density of the example disk drive.

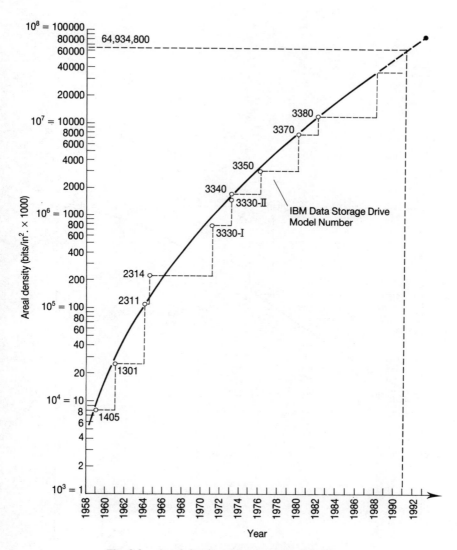

Fig. 3.3. Areal density of the example disk drive.

(13)   Linear length of the innermost track

$$G_I = \pi D_I = \pi \times 2 = 6.28 \text{ inches.} \tag{3.47}$$

(14)   Linear length of the outermost track

$$G_O = \pi D_O = \pi \times 3.74 = 11.75 \text{ inches.} \tag{3.48}$$

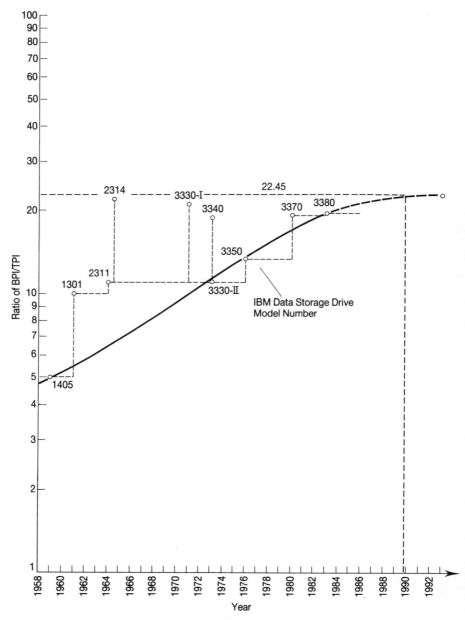

Fig. 3.4. Ratio B/T for the example disk drive.

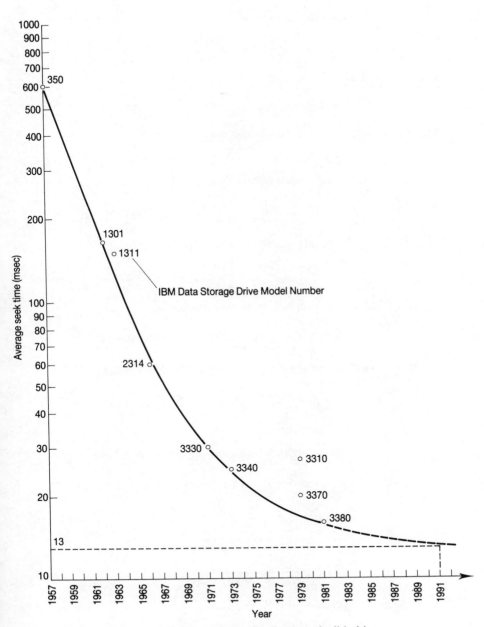

Fig. 3.5. Average seek time for the example disk drive.

(15)   Radial width of the data band

$$H = \frac{D_O - D_I}{2} = \frac{3.74 - 2}{2} = 0.87 \text{ inches.} \tag{3.49}$$

(16)   Track pitch

$$P_T = \frac{1}{T} = \frac{1}{1,700} = 588 \text{ microinches.} \tag{3.50}$$

(17)   Total number of tracks per disk side

$$K = TH = 1,700 \times 0.87 = 1479 \text{ tracks/side.} \tag{3.51}$$

(18)   "Alternate ones" fundamental frequency at the R/W head

$$1F = \frac{\Delta_B}{4(DR)} = \frac{24 \times 10^6}{4 \times 1.5} = 2.4 \text{ MHz.} \tag{3.52}$$

(19)   "All ones" fundamental frequency at the R/W head

$$2F = \frac{\Delta_B}{2(DR)} = \frac{24 \times 10^6}{2 \times 1.5} = 5.8 \text{ MHz.} \tag{3.53}$$

(20)   Data cell at the interface to the controller

$$T_C = W_{FB} = \frac{1}{\Delta_B} = \frac{1}{24 \times 10^6} = 41.67 \text{ nanosecs.} \tag{3.54}$$

(21)   Half-period of the lowest frequency at the R/W head

$$T_{FL} = (k + 1) \times \frac{m}{n} \times T_C = (7 + 1) \times \tfrac{2}{4} \times 41.67 = 4 \times 41.67$$

$$= 167 \text{ nanosecs.} \tag{3.55}$$

(22)   Half-period of the highest frequency at the R/W head

$$T_{FH} = (d + 1) \times \frac{m}{n} \times T_C = (2 + 1) \times \tfrac{2}{4} \times 41.67 = 1.5 \times 41.67$$

$$= 62.5 \text{ nanosecs.} \tag{3.56}$$

(23)   Lowest frequency at the R/W head

$$F_L = \frac{1}{2T_{FL}} = \frac{1}{2 \times 167 \times 10^{-9}} = 3 \text{ MHz.} \tag{3.57}$$

(24)   Highest frequency at the R/W head

$$F_H = \frac{1}{2T_{FH}} = \frac{1}{2 \times 62.5 \times 10^{-9}} = 8 \text{ MHz.} \tag{3.58}$$

(25)   Minimum spacing between flux changes

$$W_F(\text{minimum}) = T_{FH} = 62.5 \text{ nanosecs.} \tag{3.59}$$

(26)   Maximum spacing between flux changes

$$W_F(\text{maximum}) = T_{FL} = 167 \text{ nanosecs.} \tag{3.60}$$

(27)   Half-space between bits at the controller interface

$$T_W = \frac{1}{2\Delta_B} = \frac{1}{2 \times 24 \times 10^6} = 20.83 \text{ nanosecs.} \tag{3.61}$$

(28)   Unformatted capacity per disk side in bits

$$C_B = K \times N_B = 1{,}479 \times 240{,}000 = 355 \times 10^6 \text{ bits.} \tag{3.62}$$

(29)   Unformatted capacity per disk side in bytes

$$C_{BY} = \frac{C_B}{8} = \frac{355 \times 10^6}{8} = 44 \times 10^6 \text{ bytes.} \tag{3.63}$$

(30)   Unformatted capacity per drive in bits and in bytes. To obtain these two quantities, it becomes necessary to know the number of data surfaces (DS) and then make use of Eqs. (3.32) and (3.33). Consequently, let us assume various practical numbers for DS and prepare a table such as the one shown below:

Table 3.1.
Capacity of the Example Disk Drive.

DS	C_T (megabits)	C_{TY} (megabytes)
3	1,065	133
4	1,420	177
5	1,775	222
6	2,130	266
7	2,485	310
8	2,840	355
9	3,195	399

(31)   Linear speed at the innermost track

$$L_I = (\pi D_I) \times S = (\pi \times 2) \times 100 = 628 \text{ inches/sec.} \tag{3.64}$$

(32)   Linear speed at the outermost track

$$L_O = (\pi D_O) \times S = (\pi \times 3.74) \times 100 = 1{,}175 \text{ inches/sec.} \tag{3.65}$$

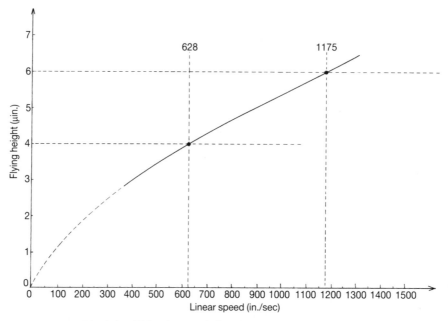

Fig. 3.6. Slider characteristics of the example disk drive.

(33) Flying height. The Read/Write head flying height is determined by the slider. Consequently, for the magnetic recording densities considered in this example, we should expect a slider with the approximate characteristics shown in Fig. 3.6. That is, it is possible to say that, approximately, the R/W head will fly from 4 microinches at the innermost track, to 6 microinches at the outermost track.

3.3. GRAPH

Equation (3.13) is so basic as a relationship between three parameters that it will be repeated here:

$$\Delta_B = S \times N_B. \tag{3.66}$$

It is possible to choose a specific value for S, in which case Eq. (3.66) can be plotted in a log–log paper giving a straight line with a slope of 45°.

At the same time, from Eq. (3.10), we have that

$$N_B = B \times (\pi D_I), \tag{3.67}$$

and from Eq. (3.8),

$$B = (DR) \times F. \tag{3.68}$$

Substituting (3.67) and (3.68) in (3.66)

$$\Delta_B = S \times (\pi D_I) \times (DR) \times F, \tag{3.69}$$

but we have that

$$(DR) = 1.5 \quad \text{and} \tag{3.70}$$

$$D_i = 2 \text{ inches.} \tag{3.71}$$

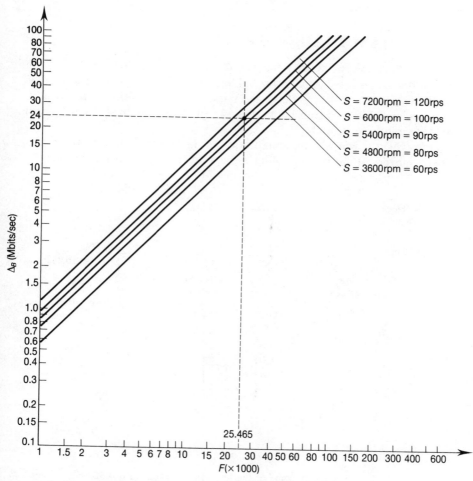

Fig. 3.7. Data rate constraints imposed by the rotational (spindle) motor speed.

Substituting (3.70) and (3.71) in (3.69) results in

$$\Delta_B = (9.424778) \times S \times F. \qquad (3.72)$$

Equation (3.72) has been plotted in Fig. 3.7. for several common values of S. This is an extremely useful graph.

The two parameters:

$$S = 100 \text{ RPS}, \quad \text{and}$$

$$F = 25,465 \text{ flux changes/inch}$$

have been entered in Fig. 3.7.

3.4. SYSTEM DIAGRAM

Figure 3.8. shows the system diagram for a disk drive. This block diagram has not changed substantially throughout the years. However, the specific implementations have undergone radical improvements, specifically, from vacuum tubes to transistors, from transistors to hybrid circuits, from hybrid circuits to ICs, etc.

However, the specific implementation of each block falls into the realm of specialized circuit design, which is beyond the scope of this book.

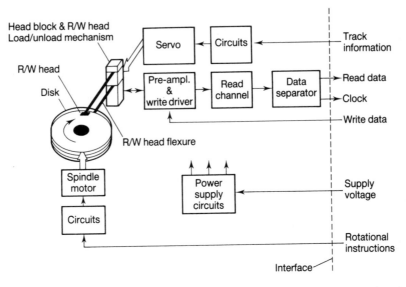

Fig. 3.8. System diagram. In some of the less costly versions of disk drives, the servo has been replaced by a stepping motor.

3.5. SUMMARY

This chapter has included a collection of formulas that define the performance of DASDs. A specific example has served to illustrate the usage of those formulas. Also, a log–log graph incorporated various parameters that are important to DASDs, thus demonstrating their interplay. This interdependence can be interpreted as a constraint (or limitation) on DASD performance. However, in actual practice, this interrelationship serves to indicate the form in which the various parameters affect each other, thus offering a glimpse of the methods that can be used to optimize whichever parameter is desired.

4

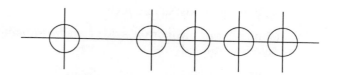

Capacity
Increase by
Data Banding

Throughout the years, DASD capacity has been increasing, mainly because of the specialized improvement of components like redesigned Read/Write heads, better magnetic materials and substrates, the addition of more platters, etc. This chapter covers only one such improvement, namely, the organization of the data in discrete bands (or zones). This approach affects not only the components but the system and programming. Although a great deal of effort has been devoted to making the consequences of this approach transparent to the user, it is imperative that the user understands the problems that this approach entails. Armed with this knowledge, the user can take advantage of increased DASD capacity by providing a judicious programming system.

4.0. BANDING

This is a technique whereby the tracks on a disk surface are grouped as concentric circles having at least one parameter in common. Without loss of generality, let us consider for the time being that this common parameter is the data rate Δ_B. To simplify the mathematical calculations let us assume that

$$B = F \tag{4.1}$$

where

$$B = BPI = \text{bit density in bits/inch,} \quad \text{and}$$

$$F = FCI = \text{flux density in flux changes/inch,}$$

thus giving

$$DR = \text{density ratio} = \frac{B}{F} = \frac{B}{B} = 1, \tag{4.2}$$

in which case

$$\Delta_F = \Delta_B = \text{data rate in bits/sec.}$$

The nomenclature to be used is as follows:

$T = \text{TPI} = \text{track density in tracks/inch.}$

$A = \text{areal density in bits/square inch.}$

$C = \text{total number of bits per disk side } capable \text{ of being obtained with the technology at hand.}$

$N = C_b = \text{unformatted capacity/disk side in bits.}$

$D = \text{outermost track diameter, for the case of a single band, only and exclusively, } D = D_O.$

It is possible to state that for a given recording capability, the maximum areal density is given by

$$A_{max} = B \times T. \tag{4.3}$$

4.1. SINGLE BAND

Equation (4.3) is quite applicable in the case of tape drives where the bits are deposited in a square fashion. However, in the case of disk drives, the bits are laid out in a radial fashion. That is, if for a particular band we wish to maintain a constant data rate, the maximum bit density (along a circumference) is dictated by the innermost and smallest track. This results in the anomaly that in the outermost and largest track. the bits are quite separate. This constant data rate is accomplished not only at the expense of losing possible capacity but also by maintaining a constant rotational speed for that band, thus obtaining exactly the same number of bits in each track, in spite of the fact that the outside track is capable of

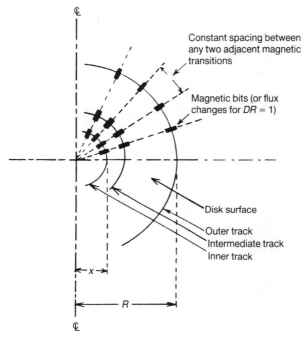

Constant spacing between
any two adjacent magnetic
transitions

Magnetic bits (or flux
changes for $DR = 1$)

Disk surface
Outer track
Intermediate track
Inner track

Fig. 4.1. Total number of bits constrained by the innermost track radius, single band. To maintain a constant data rate, all tracks have exactly the same number of bits.

supporting many more bits. This is illustrated in Fig. 4.1. for the case of a single band.

A question comes up immediately: What is the optimum radius for the inner track? In other words, what is the radius for the inner track that would offer the maximum capacity possible? Let us call

R_O = outer track radius,   and

x = inner track radius (the independent variable to be optimized).

Simple arithmetic shows that the total number of bits N is given by

$$N = 2\pi BT(R_O x - x^2), \tag{4.4}$$

which is the equation for a parabola.

It is reasonable to expect an equation such as (4.4) with a positive term and a negative term. The positive term is obtained by decreasing the innermost radius, thus giving more bits which increase the capacity.

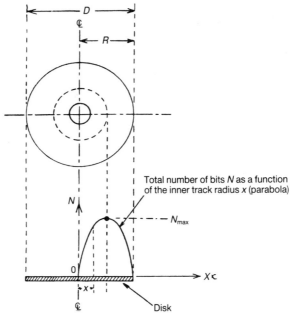

Fig. 4.2. Total number of bits per disk side as a function of the innermost track radius and for a constant data rate, single band. To maintain a constant data rate, all tracks have exactly the same number of bits. Nomenclature.

B = Bits per inch in the inner track.
T = Tracks per inch.
R_o = Outer track radius.
D_o = Outer track diameter = $2R = D$.
χ = Inner track radius (the independent variable).
$2\pi B\chi$ = Number of bits per track.
$T(R_o - \chi)$ = Number of tracks.
N = Total number of bits per surface
$\quad = (2\pi B x) \cdot T(R_o - x) = 2\pi BT(R_o\chi - \chi^2).$
$N_{max} = \dfrac{BT\pi D^2}{8}$

However, by doing this, and to maintain the data rate constant, the outer tracks are forced to have the bits more separated than before, thus losing capacity.

To obtain the maximum value for N in Eq. (4.4) let us take the first derivative and equate it to zero. This results in

$$\chi = \frac{R_O}{2}. \tag{4.5}$$

Substituting (4.5) in (4.4)

$$N_{max} = \frac{BT\pi}{2} R_O^2,$$

and for

$$D = D_O = \text{outermost track diameter} = 2R_o$$

$$N_{max} = \frac{BT\pi}{8} D^2. \tag{4.6}$$

This is shown graphically in Fig. 4.2.

It is seen that, although we have a total disk surface area of

$$A = \frac{\pi D^2}{4},$$

and we have a capability of

$$C = (BT)\frac{\pi D^2}{4} \text{ bits,} \tag{4.7}$$

even in the optimum case of Eq. (4.6), we have an efficiency of

$$\eta = \frac{N_{max}}{C}(100) = \frac{BT\pi D^2/8}{BT\pi D^2/4}(100) = 50\% \tag{4.8}$$

which is extremely low. Consequently, several methods have been considered to increase that efficiency.

4.2. MULTIPLE BANDS

One of the most obvious methods consists of maintaining the same linear density B (for each innermost track of each band), as shown in Fig. 4.3. Of course, when this is done, if we wish to maintain a constant data rate Δ_B, each track could have its own rotational speed.

To treat the problem in a general way, let us divide the disk area in concentric circles, thus giving n bands, where n can vary from 1 (for the case we use today), to the total number of tracks. Let us assume that

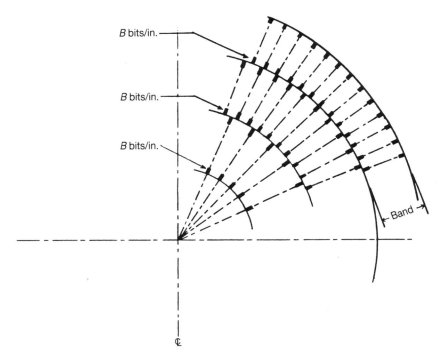

Fig. 4.3. Bits at the outside tracks constrained by the innermost track diameter of each band, multiple bands.

these bands all have the same pitch

$$\frac{D - y}{2n}$$

where

$$y = \text{diameter of the innermost track.}$$

This is shown in Fig. 4.4.

With this configuration, the total number of bits N per disk surface is given by

$$N = BT\pi \left[\frac{D - y}{2n}\right]\left\{y + \left[y + \frac{(D - y)}{n}\right] + \left[y + \frac{2(D - y)}{n}\right]\right.$$

$$\left. + \left[y + \frac{3(D - y)}{n}\right] + \ldots + \left[y + \frac{(n - 1)(D - y)}{n}\right]\right\};$$

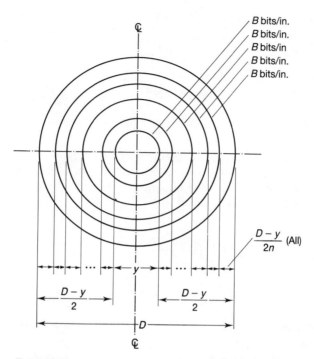

Fig. 4.4. Equally spaced bands with the same bit density B in each of the innermost tracks of each band.

$$N = BT\pi \left[\frac{D-y}{2n}\right]\left\{ny + [1+2+3+\ldots+(n-1)]\frac{(D-y)}{n}\right\};$$

$$N = BT\pi \left[\frac{D-y}{2n}\right]\left[ny + \frac{n(n-1)(D-y)}{2n}\right];$$

$$N = \frac{BT\pi}{4n}[(n-1)D^2 - (n+1)y^2 + 2Dy]. \qquad (4.9)$$

To obtain the maximum number of bits let us differentiate Eq. (4.9) and equate it to zero

$$\frac{dN}{dy} = \frac{BT\pi}{4n}[-2(n+1)y + 2D] = 0,$$

which gives

$$y = \frac{D}{n+1}. \qquad (4.10)$$

Substituting (4.7) and (4.10) in (4.9), the optimum number of bits/disk surface is then

$$N_{opt} = \left[\frac{n}{n+1}\right] C \tag{4.11}$$

and the surface efficiency is

$$\eta = \frac{N_{opt}}{C} \times 100 = \left[\frac{n}{n+1}\right] \times 100 \tag{4.12}$$

Consequently, the optimum surface partitioning is shown in Fig. 4.5. Several examples are shown graphically in Fig. 4.6. a rather significant figure of merit is given by the percentage increase in capacity with respect to the case $n = 1$. This is given by

$$I_n = \frac{\eta_n - \eta_1}{\eta_1}(100) = \frac{\frac{n}{n+1} - \frac{1}{2}}{\frac{1}{2}}(100), \quad \text{and}$$

$$I_n = \left[\frac{n-1}{n+1}\right] \times 100\%. \tag{4.13}$$

Table 4.1. shows a tabulation of some of the most important parameters.

For the ultimate case, namely, when the number of bands is equal to the total number of tracks,

$$n = \frac{TD}{2}. \tag{4.14}$$

Substituting (4.14) in our equations we have that

$$N_{opt} = \left[\frac{n}{n+1}\right] C = \left[\frac{\frac{TD}{2}}{\frac{TD}{2}+1}\right] C,$$

and for

$$\frac{TD}{2} \gg 1, \qquad N_{opt} \approx C.$$

Diameter of the innermost track:

$$\frac{D}{n+1} = \frac{D}{\frac{TD}{2}+1} \approx \frac{2}{T}.$$

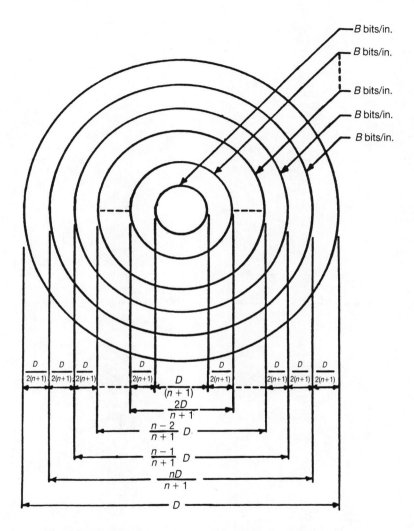

Fig. 4.5. Optimum surface partitioning for bands of equal width.

Radial distance separating each zone (pitch):

$$\frac{D}{2(n+1)} = \frac{D}{2\left[\dfrac{TD}{2}+1\right]} \approx \frac{1}{T}.$$

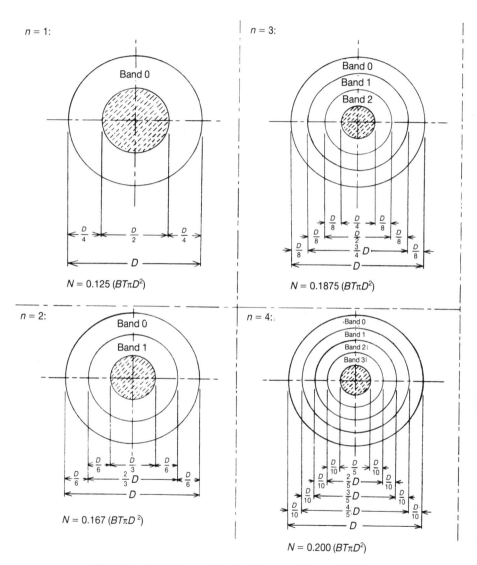

Fig. 4.6.   Several examples of optimum surface partitioning.

n = 5:

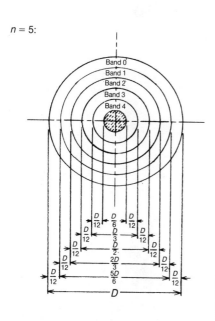

$N = 0.208 (BT\pi D^2)$

n = 6:

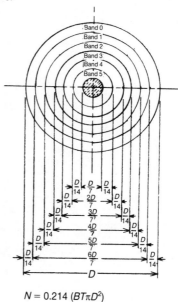

$N = 0.214 (BT\pi D^2)$

Notes:

1. For the ultimate case, that is, for $n = \dfrac{TD}{2}$, then
   $N = 0.250 (BT\pi D^2)$

2. Pitch $= \dfrac{D}{2(n + 1)}$

3. $D =$ Diameter of the disk outermost track.

Fig. 4.6. (*Continued.*)

Table 4.1.
Tablulation of Some of the Most Important Parameters.

Number of bands n	Optimum No. of bits N	Diameter of innermost track	Pitch	Efficiency η in %	Increase (%)
1	$\frac{1}{2}C$	$\frac{1}{2}d$	$\frac{1}{4}D$	50	0
2	$\frac{2}{3}C$	$\frac{1}{3}D$	$\frac{1}{6}d$	66.67	33.33
3	$\frac{3}{4}C$	$\frac{1}{4}D$	$\frac{1}{8}d$	75	50
4	$\frac{4}{5}C$	$\frac{1}{5}d$	$\frac{1}{10}D$	80	60
5	$\frac{5}{6}C$	$\frac{1}{6}D$	$\frac{1}{12}D$	83.33	66.67
6	$\frac{6}{7}C$	$\frac{1}{7}D$	$\frac{1}{14}D$	85.71	71.43
7	$\frac{7}{8}C$	$\frac{1}{8}D$	$\frac{1}{16}D$	87.5	75
8	$\frac{8}{9}C$	$\frac{1}{9}D$	$\frac{1}{18}D$	88.89	77.78
$\vdots$	$\vdots$	$\vdots$	$\vdots$	$\vdots$	$\vdots$
n	$\left[\dfrac{n}{n+1}\right]C$	$\left[\dfrac{1}{n+1}\right]D$	$\left[\dfrac{1}{2(n+1)}\right]D$	$\left[\dfrac{n}{n+1}\right]\times 100$	$\left[\dfrac{n-1}{n+1}\right]\times 100$
$\vdots$	$\vdots$	$\vdots$	$\vdots$	$\vdots$	$\vdots$
$\dfrac{TD}{2}$	C	$\dfrac{2}{T}$	$\dfrac{1}{T}$	100	100

Surface efficiency:

$$\eta = \left[\frac{n}{n+1}\right]\times 100 = \left[\frac{\dfrac{TD}{2}}{\dfrac{TD}{2}+1}\right]\times 100 \approx 100\%.$$

Increase in capacity:

$$I_n = \left[\frac{n-1}{n+1}\right]\times 100 = \left[\frac{\dfrac{TD}{2}-1}{\dfrac{TD}{2}+1}\right]\times 100 \approx 100\%, \quad (\text{for } TD \gg 1).$$

Figure 4.7. is a graph of number of bands n versus the optimum number of bits N_{opt}. Several comments seem pertinent:

(A) The process is asymptotic: That is, the greatest amount of capacity gain takes place for a smaller number of bands; thereafter, for larger number of bands, the gains start to become almost negligible. This is a typical case of "diminishing returns."

(B) As the number of bands become larger, the diameter of the innermost track becomes so small that it is impossible to locate a hub, obviously, a most impractical situation.

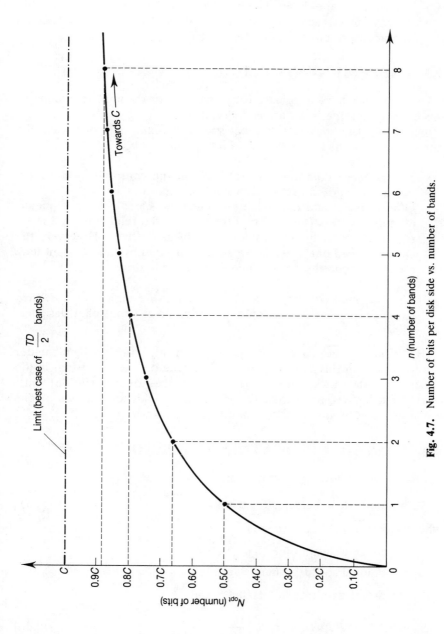

Fig. 4.7. Number of bits per disk side vs. number of bands.

(C) Fortunately, the inner bands have small capacity. That is, the increase in capacity they afford is so small that it might become impractical to pay the price for the troubles they cause.

4.3. IMPLEMENTATIONS

Several methods have been used (or proposed) for the implementation of this binding concept. They are as follows:

(A) Keep the spindle rotational speed constant, in which case the rate of the data bits for each band is different.

(1) Accept those different data rates Δ_B in the controller and offer a range (as opposed to a single value) of data rates.
(2) Offer a single data rate value but affect the speed change by means of shift registers (buffers, or FIFOs). For the read case, the clock rate to load to buffers are dictated by the drives. However, the clocks to read out of the buffers are controlled by this constant data rate. The opposite is true when writing in the drive.

The two methods mentioned above make life easier for the drive designer but they impose quite a burden on the controller designer. Thus:

(B) Change the rotational speed according to the band under consideration. In this case, in order to keep the data rate constant in the drive itself, the tracks with the smaller number of bits (inside) must rotate at a higher speed than the tracks with a higher number of bits (outside). A very simplified system diagram is shown in Fig. 4.8.

4.4. VARIABLE ROTATIONAL SPEED

Let us use the following nomenclature:

$$D_k = (\text{inner track diameter for band } k) = \frac{(n-k)D}{n+1}, \qquad (4.15)$$

where: $k = 0, 1, 2, 3, \ldots, (n-1)$,

$L_k = $ inner track length for band $k = \pi D_k$

$S_k = $ rotational speed for band k in revolutions/sec,

$\Delta_B = $ constant data rate in bits/sec,   and

$$\Delta_B = BS_k L_k = BS_k \pi D_k = \pi BS_k \left[\frac{(n-k)D}{(n+1)} \right] \qquad (4.16)$$

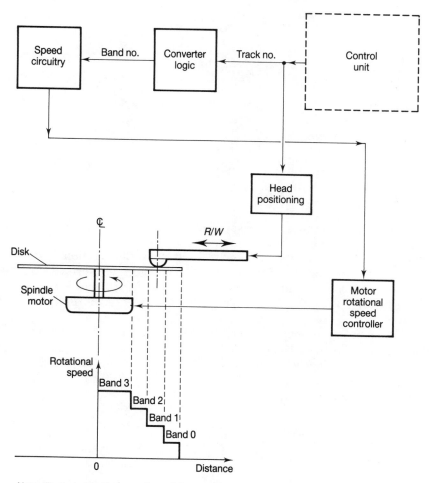

Note: Illustrated for the case of $n = 4$ (four bands)

Fig. 4.8. System diagram and speed for each band when the data rate is kept constant. Illustrated for the case of $n = 4$ (four bands).

therefore, the rotational speed for band k is given by

$$S_k = \frac{\Delta_B}{\pi B}\left[\frac{(n+1)}{(n-k)D}\right]. \tag{4.17}$$

There are many ways to implement this method and, undoubtedly, many others will appear in the literature. As an example, we are going to

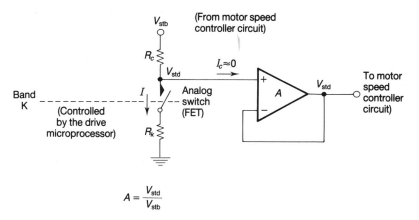

$$A = \frac{V_{std}}{V_{stb}}$$

Fig. 4.9. Method of spindle motor speed control by means of voltage gains.

illustrate only one of them, namely, controlling the rotational speed of the spindle by means of a voltage-to-frequency converter. For this method, the spindle motor speed is not constant, rather, it is variable (asjustable) and very much dependent on the voltage applied to the motor.

Consequently, we must find a method to control the voltage applied to the spindle motor. This can be accomplished rather easily by means of an operational amplifier whose gain is controlled by means of two resistors, as shown in Fig. 4.9.

Referring to that figure, we can say

$$I = \frac{V_{stb}}{R_c + R_k}.$$

Also

$$I = \frac{V_{std}}{R_k}.$$

Therefore,

$$\frac{V_{stb}}{R_c + R_k} = \frac{V_{std}}{R_k},$$

or

$$\frac{V_{std}}{V_{stb}} = \frac{R_k}{R_c + R_k}. \tag{4.18}$$

Let us call

$$A = \frac{V_{std}}{V_{stb}}. \tag{4.19}$$

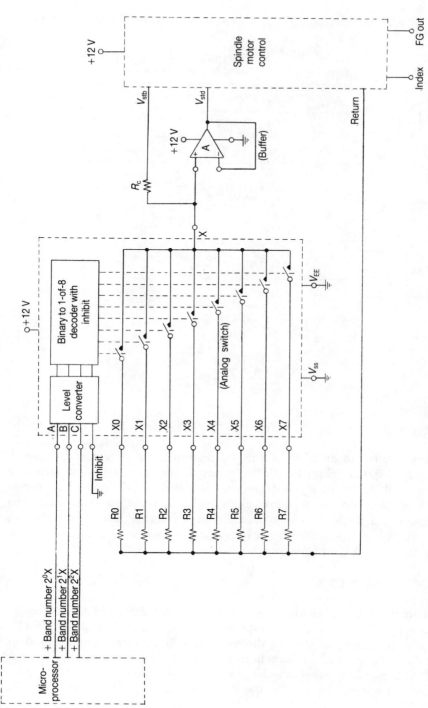

Fig. 4.10. Circuit implementation of several motor speeds by means of analog switches. Illustrated for the case of $n = 8$ (eight bands).

95

Table 4.2.
Table for a Typical Analog Multipler/Demultiplexer.

	Controls inputs			
		Select		ON
Inhibit	C	B	A	Switch
0	0	0	0	$\chi\phi$
0	0	0	1	$\chi 1$
0	0	1	0	$\chi 2$
0	0	1	1	$\chi 3$
0	1	0	0	$\chi 4$
0	1	0	1	$\chi 5$
0	1	1	0	$\chi 6$
0	1	1	1	$\chi 7$
1	χ	χ	χ	None

Substituting (4.19) in (4.18)

$$A = \frac{R_k}{R_c + R_k},$$

which gives

$$R_k = \left[\frac{A}{1-A}\right] \times R_c. \tag{4.20}$$

It is possible to make R_c a constant resistor, thereafter it is possible to insert different values for R_k by means of analog switches, as shown in Fig. 4.10. Those analog switches are controlled by the drive microprocessor which inserts the various values for R_k according to Table 4.2. The ratio in Eq. (4.20) can be obtained from the motor manufacturer's data on its voltage-to-frequency converter (that controls the motor speed). A curve of this ratio is shown in Fig. 4.11.

4.5. SUMMARY

Many methods have been proposed (and implemented) that increase the capacity of DASDs. This chapter has covered one such method, namely, the arrangement of data in discrete bands (or zones). One method of making the method transparent to the user has been included above, that is, the circuitry necessary for an adjustable speed rotational motor. Another approach (that has also been used) consists of an array of shift

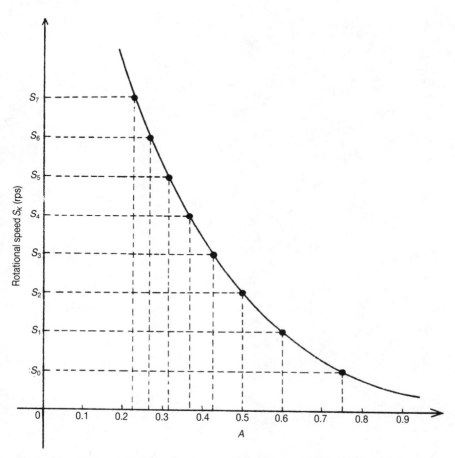

Fig. 4.11. Voltage-to-frequency converter curve for the spindle rotation motor. Illustrated for the case of eight ($n = 8$) bands.

registers in the disk controller. In any case, the equations offered previously will serve to calculate the various rates needed in the data processing environment.

5

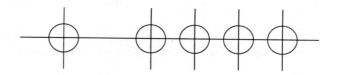

Interaction between Platters and Read/Write Heads

The transducers (Read/Write (R/W) heads) fly at very close proximity to the platter surfaces that rotate at rather high speeds. This close proximity of two hard surfaces creates a great deal of friction, and the corresponding interaction becomes a specialty of mechanical engineering named tribology. The situation is aggravated by the fact that each R/W head is mounted on a device (slider) whose surface is so designed that its profile ensures this close proximity. The sliders are attached to the header by means of metallic arms (flexures) whose stiffness must be accurately maintained. This chapter introduces some of the most important parameters along with their interaction and interdependence.

5.0. TRIBOLOGY

Tribology is the science and technology of interacting surfaces in relative motion and the practices relating thereto. An object moves with relation to another object that may or may not be stationary but moves at a slower speed than the object under consideration. For example, in the case of an airplane, the object (airplane) moves with respect to another object (the earth) and it is possible to determine its speed (with respect to ground) and also its flying height. In the case of disk drives, the object (R/W head) also flies (but it remains stationary) while the disk moves (by rotation) underneath.

Also, the R/W head flying heights are so small that there is a tremendous amount of friction between the head and the disk surface. Therefore, the laws of tribology are very applicable to this case, and in particular, to the design of the slider profile.

5.1. LINEAR SPEED

When we talk about flying "speed" in disk drives, we are really referring to the speed of the rotating disk, not the speed of the R/W head, which remains stationary. Thus, we must expect the linear speed to be a function of track dimension and disk rotational speed.

In the case of a single band, equations are listed below:

For the innermost track (which offers the minimum linear speed):

$$L_I = (\pi D_I) \times S. \qquad (5.1)$$

For the outermost track (which offers the maximum linear speed):

$$L_O = (\pi D_O) \times S. \qquad (5.2)$$

But we have that

$$S = \frac{\Delta_B}{(\pi D_I) \times B}. \qquad (5.3)$$

Substituting (5.3) in (5.1)

$$L_I = \frac{\Delta_B}{B}. \qquad (5.4)$$

But it was determined that

$$B = (DR) \times F, \qquad (5.5)$$

and for the RLL $(2, 7)$ code, it was determined that

$$(DR) = 1.5. \qquad (5.6)$$

Substituting (5.6) and (5.5) in (5.4)

$$L_i = \frac{\Delta_B}{1.5 \times F}. \qquad (5.7)$$

If we assume a fixed value for Δ_B, the plot of Eq. (5.7) in a log–log

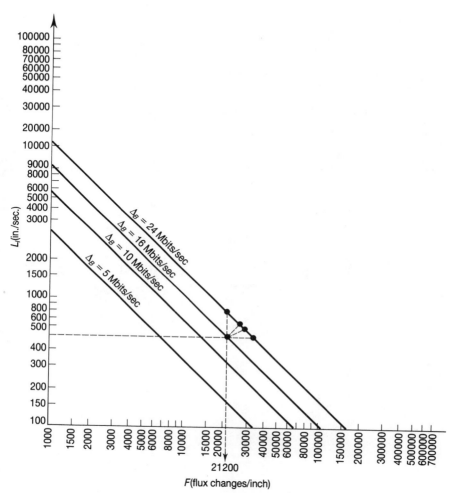

Fig. 5.1. Linear speed constraints imposed by the data rate.

paper will result in a straight line with a −45° slope. This is shown in Fig. 5.1.

Example. Let us consider the following parameters:

$$D_I = 2 \text{ inches}; \tag{5.8}$$

$$D_O = 3.74 \text{ inches}; \tag{5.9}$$

$$S = 80 \text{ RPS}; \tag{5.10}$$

$$F = 21{,}200 \text{ flux changes/inch}; \quad \text{and} \tag{5.11}$$

$$\Delta_B = 16 \text{ million bits/second}. \tag{5.12}$$

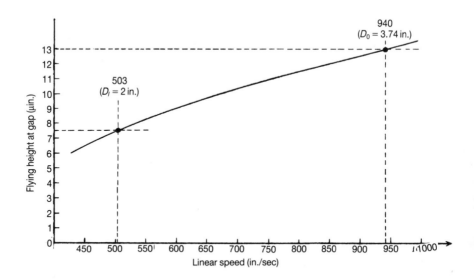

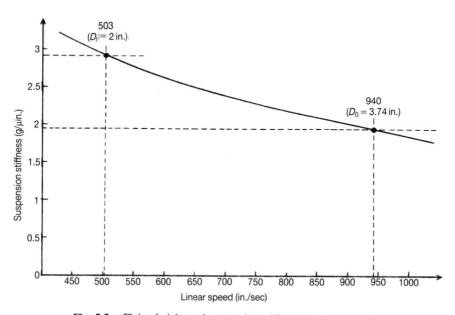

Fig. 5.2. Flying height and suspension stiffness vs. linear speed.

Applying Eq. (5.1)

$$L_I = (\pi \times 2) \times 80 = 502.65 \text{ inches/sec.} \qquad (5.13)$$

The point given by (5.11), (5.12), and (5.13) has been shown in Fig. 5.1. Applying Eq. (5.2)

$$L_O = (\pi \times 3.74) \times 80 = 940 \text{ inches/sec.} \qquad (5.14)$$

Consequently, for this particular case, we can expect the linear speed to vary from 503 inches/sec to 940 inches/sec.

5.2. FLYING HEIGHT

The number of bits per inch B affects disk capacity, and B is primarily a function of head flying height: the smaller the flying height, the larger the

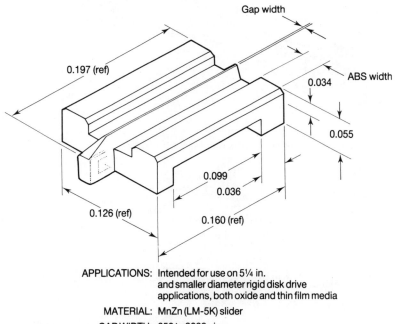

APPLICATIONS: Intended for use on 5¼ in. and smaller diameter rigid disk drive applications, both oxide and thin film media

MATERIAL: MnZn (LM-5K) slider

GAP WIDTH: 650 to 2000 µin. (0.0165 to 0.0508 mm) ±10%

GAP LENGTH: 40±µin. (1±0.2 micron)

GAP DEPTH (THROAT HEIGHT): 0.0007±0.0003 in. (0.0178±0.0074 mm)

AIR BEARING SURFACE (ABS): 0.027 to 0.031±0.0005 in. (0.686 to 0.787±0.013 mm)

CENTER RAIL ANGLE: 60±2°

SLIDER WEIGHT: 0.066±0.005 g

Fig. 5.3. Dimensions of a commercially available slider. Reproduced with permission from National Micronetics, Inc.

value of *B*. For instance, the limitation was 120 microinches flying height for the 40–inch diameter disk. This height is projected to be approximately 12 microinches for the 14–inch disk, and we can expect less than 4 microinches for $3\frac{1}{2}$ inch disks.

The slider is the device that allows the head to fly above the disk in such close proximity. The slider is held in place by the suspension. Figure 5.2. shows the measured characteristics of a slider and supension currently available. If we enter our calculated range of linear speeds on

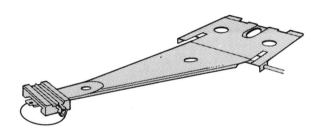

HEAD AND GIMBAL (HGA) ASSEMBLY

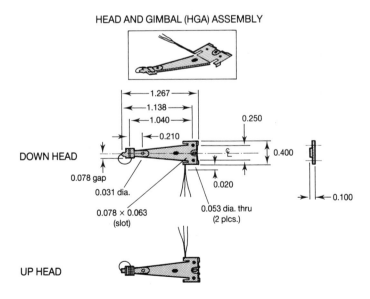

Fig. 5.4. Slider and thin-film head in a Whitney-type flexure. Reproduced with permission from National Micronetics, Inc.

these graphs, we find that the R/W head flying height would vary from 7.5 microinches (for the innermost track) to 13 microinches (for the outermost track). Also, the suspension stiffness would vary from 2.9 grams/microinch to 1.9 grams/microinch.

Figure 5.3. shows the dimensions of a commercially available slider and Fig. 5.4. shows that slider with a thin-film R/W head in a Whitney-type

95 mm TEST DESCRIPTION

THIN FILM MEDIA

TEST PARAMETERS

1F FREQUENCY:	1.25MHz
2F FREQUENCY:	2.50MHz
INSIDE RADIUS:	1.00 INCHES (25.40mm)
OUTSIDE RADIUS:	1.70 INCHES (43.18mm)
DISK SPEED:	3600 RPM
VELOCITY, INSIDE RADIUS:	377 IN/S (9.5m/s)
VELOCITY, OUTSIDE RADIUS:	641 IN/S (15.5 m/s)
PACKING DENSITY,INSIDE RADIUS:	13,262 FLUX CHANGES/IN
READ LOAD:	560 OHMS
WRITE CURRENT:	20 TO 60mA,p-p
RISE TIME:	30 NANOSECONDS AT 55mA

THIN FILM MEDIA TEST HEAD AVERAGES

GAP WIDTH:	.00132 INCH (.033mm)
GAP LENGTH:	43 MICROINCHES (1.09 MICRONS)
THROAT HEIGHT:	640 MICROINCHES (16.25 MICRONS)
FLYING HEIGHT:	13 MICROINCHES (.33 MICRONS)
AIR BEARING SURFACE (ABS):	.027 INCH (.68mm)
NUMBER OF TURNS:	20 BIFILAR

TEST MEDIA

95mm DISK 600 Oe

Fig. 5.5. Laboratory conditions for R/W measurements taken with a commercially available thin-film head. Reproduced with permission from National Micronetics, Inc.

flexure. These figures serve to indicate the dimensions needed for this density of recording. It is expected that in the future these dimensions will become even smaller.

For purposes of illustration, Fig. 5.5. shows typical laboratory conditions for R/W measurements to be taken with the previously illustrated thin-film head, slider, and suspension.

5.3. SUMMARY

The interaction between the R/W heads and the platter surfaces becomes a very complicated phenomenon when the DASD is in operation. This nearness between them creates a great deal of friction, and thus it falls in the realm of tribology.

6

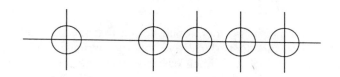

Data
Encoding for
Magnetic Recording

The binary data to be stored in a DASD is first generated by a computer. Thereafter, that data (plus clocking information) is sent to the controller. However, this data (plus clock) must be translated into a form capable of being accepted by the storage phenomenon being used. Mostly, this phenomenon is magnetic recording in which data is stored only as magnetic transitions. This chapter constitutes an introduction to that conversion (named encoding), which results in various codes. During readback from the storage medium, the opposite operation takes place (named decoding), that is, the recorded transitions must be translated into pulses capable of being understood by the computer. This dual translation can take place inside the DASD itself (by means of the appropriate circuitry), in the controller, or in the computer (by means of programming). Obviously, if this dual translation is performed automatically in the DASD (as is usually done), then the operations become transparent to the user.

6.0. CODING

There was a time when the process of choosing a code (for digital magnetic recording) was lacking in rigorousness, and therefore was surrounded by emotion, stress, and lack of objectivity. As a result, there has been a tremendous proliferation of such codes, as shown in Fig. 6.1., a veritable "tower of Babel" leading to unnecessary confusion and lack of standardization. Fortunately, this situation does not exist any longer due to the fact that all codes can be analyzed mathematically using a standard nomenclature. This subject will be covered in the next chapter. However, there are many disk drives (and tapes) that are still using some of these codes, therefore, it becomes necessary to cover them in some detail, although today, most of them can be considered obsolete from a disk drive design point of view.

6.1. REPRESENTATION

Although it was mentioned before that the magnetic recording read head offers the first derivative of the write waveform, for purposes of explanation, it is much easier to concentrate on the write waveforms (as shown in Fig. 6.1.). This approach simplifies things significantly.

A code can be represented by means of a shift register, as shown in Fig. 6.2, where the output of each flip-flop signifies whether the bit is a one or a zero. The bit information moves from left to right. That is, the incoming bits enter the shift register from the left and leave from the right. This representation is used in error correcting codes. In magnetic recording, and for generality purposes, each transition can represent either data or clock.

A code can be described also by a pictorial representation of the corresponding waveform, or as seen in an oscilloscope. This is shown in Fig. 6.3. and also in Fig. 6.1. For this case, the bit information moves also from left to right. However, in Cartesian coordinates when the abscissa is "time" (as in an oscilloscope), the "future" is to the right, and the "past" is to the left, which is exactly the opposite of what happens in a shift register representation.

It is obvious that there is a discrepancy when representing a code by shift registers or by waveforms. However, this only involves a time reversal when going from one pictorial representation to the other. Consequently, we will use this time reversal when required, but to avoid further confusion, the case under consideration will be identified.

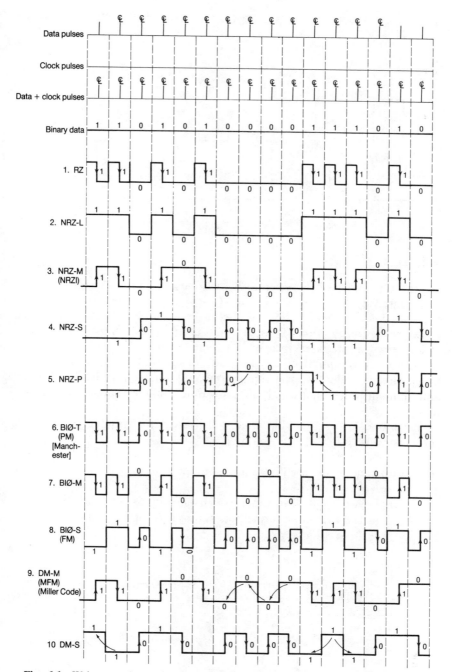

Fig. 6.1 Write waveforms for some codes that have been used in digital magnetic recording.

109

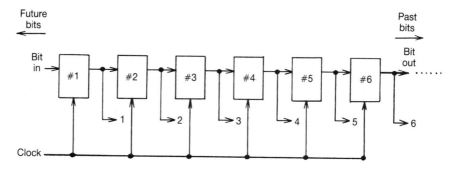

Fig. 6.2.   Shift register representation of a code.

6.2. BIT CELL

For the remainder of this chapter we will describe several codes that are represented by the write waveforms (not the read waveforms). For this purpose, let us define a bit cell that will be used later on during the waveform representation of a code. First of all, a code consists of a series of bit cells concatenated in a time sequence. Of course, ultimately, the bits will be stored physically in a portion of the disk surface. Meanwhile, let us disregard the physical dimensions and concentrate instead on the time domain only. Figure 6.4. shows a time domain representation of a single bit cell.

Figure 6.4. also shows that a bit cell possesses attributes horizontally and vertically.

(1)   Horizontally, the attributes are as follows:

(a)   Two clock transitions which encompass and define the bit. The first clock transition belongs to the bit, and the second

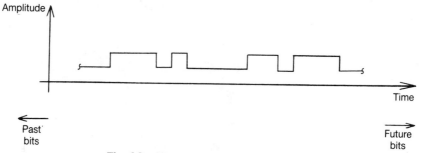

Fig. 6.3.   Waveform representation of a code.

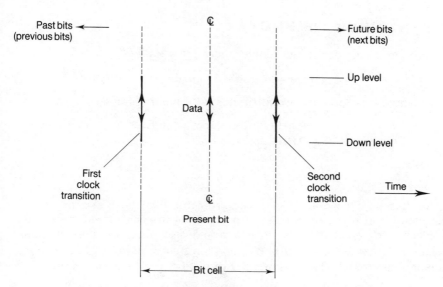

Fig. 6.4. Bit cell used in the representation of a code waveform.

transition belongs to the neighboring bit. The only exception is the last bit, for which there is no next bit, and the second clock transition is needed to terminate the bit sequence.

(b) Middle transition which is usually called the "data" transition.

(c) Any of those transitions can be up-going or down-going, therefore, for each bit we have four attributes:

> Clock up-going,
> Clock down-going,
> Data up-going,   and
> Data down-going.

(2) Vertically, there is:

> Up level,   and
> Down level.

It is possible to take into consideration the bit next to the one under consideration or the previous one. Therefore, there are four more attributes according to the bits surrounding the bit under consideration:

> Next bit equal,
> Next bit unequal,
> Previous bit equal,   and
> Previous bit unequal.

6.3. NUMBER OF CODES

According to the definitions above, there are a total of

$$m = 4 + 2 + 4 = 10 \text{ attributes.} \tag{6.1}$$

In reality, we need only two attributes for a binary number:

> Attribute for a zero, and
> Attribute for a one.

That is

$$n = 2. \tag{6.2}$$

This has now become a classical problem, namely: Take m distinct elements and make up groups of n elements in each, but disregard the order of the elements in each group and also disregard the remaining elements (that are not used). We then have combinations of m elements taken n at a time. The total number of distinct combinations is then

$$C_n^m = \binom{m}{n} = \frac{m!}{n!\,(m-n)!}. \tag{6.3}$$

Consequently, substituting (6.1) and (6.2) in (6.3), the possible number of codes with the attributes described above are:

$$C_2^{10} = \binom{10}{2} = \frac{10!}{2!\,(10-2)!} = \frac{9 \times 10}{2} = 45. \tag{6.4}$$

Of course, of those possible 45 codes, some are impractical (from an implementation viewpoint) and have never been used. Others have never been considered or even defined. Therefore, we will illustrate only some of the most commonly used codes.

At times, the clocking transitions become so important that we do not have a choice whether to use them or not, that is, for those codes there is a basic rule which says, "You *must* have a clock for every bit, however, you can disregard the direction of that clock transition."

In this case, although $n = 2$, we have lost two mandatory attributes in which case

$$m = 10 - 2 = 8. \tag{6.5}$$

Substituting (6.2) and (6.5) in (6.3), the number of self-clocking codes are:

$$C_2^8 = \binom{8}{2} = \frac{8!}{2!\,(8-2)!} = \frac{7 \times 8}{2} = 28. \tag{6.6}$$

Again, some of those codes are very impractical and only a few have been considered.

6.4. CODE DEFINITIONS

The following is a list of several code definitions (and rules).

(1)   Return to zero (RZ).

    (a)   *Always* return to the down level (baseline).
    (b)   "Zero": No transition, neither at clock time nor in the middle of the cell. Since it is required to return to the down level (baseline), this level becomes the zero.
    (c)   "ONE": Down-going transition in the middle of the cell. Therefore, it is always necessary to go up (before making a down-going transition) and this is done during the first clock transition.

(2)   Non-return to zero, level (NRZ–L).

    (a)   "Zero": Down level (baseline).
    (b)   "One": Up level.
    (c)   Sometimes it becomes necessary to go from the down level to the up level (when going from a zero to a one), and also from the up level to the down level (when going from a one to a zero). For this purpose, the first clock transition is used.

(3)   Non-return to zero, mark (NRZ–M), or non-return to zero, inverted (NRZI).

    (a)   "Zero": No transition anywhere, stay where you are, either at the down level or at the up level. This means that a zero is sometimes at a down level (baseline) and at an upper level some other times. The level is immaterial.
    (b)   "One": Transition in the middle of the cell. Sometimes the transition is up-going, sometimes it is down-going. Therefore, the direction of the transition is meaningless.
    (c)   Since the level is meaningless (for the zeros) and the direction of the transition is meaningless (for the ones), the clock transition is meaningless (for the ones); the clock transition is never used to define the code, only to encompass the bit cell.

(4)   Non-return to zero, space (NRZ–S). This is exactly the opposite definition of the previous code.

  (a)   "Zero": Transition in the middle of the cell. Sometimes the transition is up-going, sometimes it is down-going. Therefore, the direction of the transition is meaningless.

  (b)   "One": No transition. Stay where you are, either at the down level or at the up level. This means that a one is sometimes at a down level (baseline) and at the upper level some other times. The level is immaterial.

  (c)   Since the direction of the transition is meaningless (for the zeros) and the level (for the ones) is also meaningless, the clock transition is never used to define the code, only to encompass the bit cell.

(5)   Non-return to zero, previous (NRZ–P).

  (a)   Transition in the middle of the cell, only and exclusively when

$$\text{present bit} \neq \text{previous bit.}$$

  (b)   When no transition occurs in the middle of the cell, it is assumed that

$$\text{present bit} = \text{previous bit.}$$

  (c)   Since the level is meaningless (for either ones or zeros) and since the direction of the transition (for either ones or zeros) is also meaningless, the clock transition is never used to define the code but instead to encompass the bit cell.

(6)   Bi-phase, transition (BIØ–T), phase modulation (PM), Manchester code

  (a)   "Zero": Up-going transition in the middle of the cell.
  (b)   "One": Down-going transition in the middle of the cell.
  (c)   Whenever it becomes necessary to implement the proper direction of the transition, as defined in (a) and (b) above, the clock transition is used to reach the proper level (either up or down).
  (d)   Since there is always a transition in the middle of the cell (irrespective of whether the bit is a "one" or a "zero"), this transition can be used for clocking purposes. In that respect the code is "self-clocking."

(7)   Bi-phase, mark (BIØ–M).

    (a)   "Zero": No transition. This means that a zero is sometimes at a down level (baseline) and at the upper level some other times. The level is immaterial.

    (b)   "One": Transition in the middle of the cell. Sometimes the transition is up-going, sometimes it is down-going. Therefore, the direction of the transition is meaningless.

    (c)   There is always a clock transition, one for each bit. This requirement makes the code "self-clocking."

    (d)   This code is almost the same as NRZ-M (NRZI), or number (3) above. The only difference is in the clock requirement:

        BIØ–M always requires a clock transition.

        NRZ–M never uses the clock transition.

(8)   Bi–Phase, Space (BIØ–S), or also called frequency modulation (FM).

    (a)   "Zero": Transition in the middle of the cell. Sometimes the transition is up-going, sometimes it is down going. Therefore, the direction of the transition is meaningless.

    (b)   "One": No transition. This means that a one is sometimes at a down level (baseline) and at the upper level some other times. The level is immaterial.

    (c)   There is always a clock transition, one for each bit. This requirement makes the code "self-clocking."

    (d)   This code is almost the same as NRZ–S, or number (4) above. The only difference is in the clock requirement:

        BIØ–S always requires a clock transition.

        NRZ–S never uses the clock transition.

(9)   Delay modulation, mark (DM–M), MFM, Miller code, etc.

    (a)   Encode exactly as in NRZI (number (3) above), that is

        Zero = No transition.

        One = Transition in the middle of the cell.

    (b)   However, we saw that NRZI does not make use of the clock transitions at all. This presents a clocking problem in the case of a long string of zeros. In the case of ones there is no problem because the transitions in the middle of the cell can be used for clocking. However, in the case of zeros, there is no transition at all.

(c)   Since the problem is caused by a long string of zeros, look at the previous bit (some form of storage is required), if the present bit and the previous bit are both zero, insert a clock pulse between the two.

(d)   It is obvious that this code is not a modification of FM (code number (8) above), rather, it is a modification (or a clocking improvement) on the NRZI code.

(10)   Delay modulation, space (DM-S). Nobody seems to want to take credit for this code.

(a)   Encode exactly as in NRZ–S (number (4) above), that is

Zero = Transition in the middle of the cell.

One = No transition.

(b)   However, we saw that NRZ–S does not make use of the clock transitions at all. This presents a clocking problem in the case of a long string of ones. In the case of zeros, there is no problem because the transistions in the middle of the cell can be used for clocking. However, in the case of ones, there is no transition at all.

(c)   Since the problem is caused by a long string of ones, look at the previous bit (some form of storage is required), if the present bit and the previous bit are both one, insert a clock pulse between the two.

(d)   It is obvious that this code is a modification (or clocking improvement) on the NRZ–S code.

6.5. BITS PER TRANSITION

There are some applications, specifically disk drives, where it is necessary to know, a priori, the number of magnetic transitions (both, data and clock) that are needed for specific codes. This knowledge helps (sometimes) to choose a code that would offer rather good density, that is, bits/inch. For this purpose,

B = total number of bits (per some specific unit, such as inch, time interval, etc.),

F = total number of flux transitions (using the same units as above), and DR = density ratio, data bits/flux reversal.
That is:

$$DR = \frac{B}{F}. \tag{6.7}$$

Statistically, and in the limit, let us assume that the total number of bits B contains an equal number of ones and zeros, that is:

Total number of zeros: $N_0 = B/2$, and $\qquad$ (6.8)

Total number of ones: $N_1 = B/2$. $\qquad$ (6.9)

In particular, referring to Fig. 6.1 only, we can count

$$B = 16, \qquad N_0 = 8, \qquad N_1 = 8.$$

Again, it is possible to refer to Fig. 6.1. and count for each particular code the total number of flux transitions F needed to generate that code. We find the following:

(1) $F = 16$, then $DR = \frac{16}{16} = 1$.
(2) NRZ–L: $F = 9$, then $DR = \frac{16}{9}$. However, in the limit, and for $B \to \infty$, $DR = 2$.
(3) NRZI: $F = 8$, then $DR = 2$. However, in the limit, and for $B \to \infty$, $DR = 1$.
(4) NRZ–S: $F = 8$, then $DR = 2$.
(5) NRZ–P: $F = 8$, then $DR = 2$.

It is seen that, with the exception of RZ, the codes above have a rather large value of DR, which makes them very efficient, and they have been used extensively in early disk files. However, experience with their usage has shown that the problems with clocking can be formidable. For this reason, other codes have been used that offer better clocking capabilities at the expense of a lower DR. For instance:

(6) BIØ–T: $F = 22$, then $DR = 0.73$. However, for N very large, in the limit we have

$$DR = \frac{B}{B + \dfrac{B}{2}} = \frac{1}{\frac{3}{2}} = \frac{2}{3} = 0.67.$$

(7) BIØ–M: $F = 24$, then $DR = 0.67$.
(8) BIØ–S: $F = 24$, then $DR = 0.67$.

It is seen that those codes have good clocking characteristics, but the values of DR are very low (rather inefficient).

Let us consider the following definition:

d = minimum number of consecutive zeros allowed (including clock),

k = maximum number of consecutive zeros permitted (including clock),

and, always, $k > d$.

Since these transitions are related to the period of some frequency, it is possible to say that

$$\text{maximum frequency} \propto \frac{1}{d},$$

$$\text{minimum frequency} \propto \frac{1}{k},$$

and their ratio is

$$FR = \frac{\text{maximum frequency}}{\text{minimum frequency}} \propto \frac{1/d}{1/k} = \frac{k}{d}. \tag{6.10}$$

In the case of the NRZI code and for a very large sequence of zeros, it is possible to say that, in the limit

$$k = \infty, \tag{6.11}$$

substituting (6.11) in (6.10) we obtain

$$FR = \infty. \tag{6.12}$$

Equation (6.12) means that the frequency band of the read channel must be extremely wide, thus allowing high frequency noise to pass, with the consequent degradation in signal-to-noise ratio. Therefore, it becomes desirable to lower the width of the required frequency band.

For that purpose, an intermediate position (compromise) can be obtained by adding a few clocking pulses to NRZI. In that case, we obtain the "compromise" codes, which are

(9)   DM-M (MFM, Miller code, etc.)

$$DR = \frac{B}{B/2 + B/2} = 1 \quad \text{(same as NRZI),} \quad \text{and} \tag{6.13}$$

$$FR = \frac{3+1}{1+1} = \frac{4}{2} = 2 \quad \text{(much lower than infinity).} \tag{6.14}$$

(10)   DM-S. Same as above.

The search for a better code remains unabated and, very likely, this situation will continue forever. However, there are many other factors that enter into the selection of a code (much more than mere circuit design), namely, signal-to-noise ratio, error propagation, ECC, etc. The next chapter will cover a class of codes much in vogue these days.

6.6. SUMMARY

This chapter has presented some of the codes that have been used (and are still being used) that are capable of effecting the translation between the binary data pulses (and clock) into transitions capable of being written onto a magnetic surface.

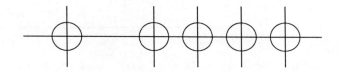

Run-Length Limited Code (2, 7)

This chapter presents an example of one of the many codes used in magnetic recording. The details included in this implementation of the Run-length Limited code $(2, 7)$ serve to illustrate the requirements that must be implemented regardless of the code being used. At the beginning of this chapter, the presentation is done in general algebraic terms so that any code at all can be evaluated (as done here for some of them) and compared with any other code. The implementation can be done with hardware (logic design), firmware or programming. All of them have been included here along with several patterns that have been found useful in detecting errors.

7.0. RUN-LENGTH LIMITED CODES

Table 7.1. illustrates several codes that have been used during the design and implementation of disk drives. Obviously, the clock transitions (and not the data transitions exclusively) are of great importance also and must be taken into consideration. Therefore, any exhaustive mathemati-

Table 7.1.

Examples of Several Codes Used in Disk Drives.

Name	Binary data		Coded representation
			data
NRZI	1		1
	0		0
			No clock
			Data
FM	1		11
	0		01
			Clock (Always)
			Data
MFM	1		10
	0[Followed by 1]		00
	0[Followed by 0]		01
			Clock (only when needed)

cal treatment of codes must give equal importance to any transition, whether it is data or clock (a transition is a transition, is a transition, is a . . .).

7.1. MATHEMATICAL EVALUATION

Let us consider the following nomenclature:

d = Minimum number of consecutive zeros allowed (including clock).

k = Maximum number of consecutive zeros permitted (including clock).

m = Minimum number of data bits to be encoded.

n = Number of code bits (including clock) for each of the m data bits.

r = Number of different word lengths in a variable length code.

P = Error propagation. The maximum number of additional bits of error resulting from one code bit in error.

DR = Density Ratio. Data bits per flux reversal:

$$DR = \frac{\text{bits per inch } (BPI)}{\text{flux changes per inch } (FCI)} = (d + 1) \times \left(\frac{m}{n}\right).$$

FR = Frequency Ratio:

$$FR = \frac{\text{maximum time between transitions}}{\text{minimum time between transitions}} = \frac{k + 1}{d + 1}.$$

w = Detection window expressed as a percentage of a data bit cell

$$w = \left(\frac{m}{n}\right) \times 100.$$

The codes that result can be called "Run-Length Limited." The full designation would be:

$$\text{RLL}(d, k, m, n; r).$$

However, this has been shortened to

$$\text{RLL}(d, k)$$

Table 7.2.

Mathematical Calculations Performed for Several Codes.

Code name	Where used	d	k	m	n	r	DR	FR	w
NRZI	Early disk drives	0	∞	1	1	1	1.0	∞	100%
FM	Floppy disks	0	1	1	2	1	0.5	2.0	50%
GCR	IBM 3420 tape	0	2	4	5	1	0.8	3.0	80%
8:9		0	3	8	9	1	0.89	4.0	89%
MFM	IBM 3330 (plus everybody)	1	3	1	2	1	1.0	2.0	50%
3PM	ISS	2	11	3	6	1	1.5	4.0	50%
RLL (1, 8)	APS drives	1	8	2	3	3	1.33	4.5	67%
RLL (2, 7)	IBM 3370 (plus everybody)	2	7	2	4	3	1.5	2.67	50%
RLL (1, 5)	(Nowhere)	1	5	6	10	1	1.2	3.0	60%
RLL (1, 7)	Quantum Q160	1	7	7	11	1	1.27	4.0	64%
RLL (1, 6)	(Nowhere)	1	6	2	3	5	1.33	3.5	67%

With this nomenclature, it becomes possible to calculate those parameters for any code past, present, or future. This way it becomes possible to prepare a table such as that shown in Table 7.2.

Also, it is possible to obtain the write waveform (for arbitrary data) using any such codes, such as the RLL (2, 7) shown in Fig. 7.1., although at this time we do not yet know the corresponding NRZ data. This last piece of information will be furnished by the encode/decode table being implemented.

From an information theory contents point of view, all those codes are equivalent (Macintosh, 1979, 1980). Today, and with the full benefit of 20/20 hindsight, it seems almost obvious that it should be that way. However, it is extremely gratifying to be certain that this point has been proven rigorously and mathematically.

Therefore, other than circuit design complexity, some criteria must be used in the code selection process. Let us choose a code with the

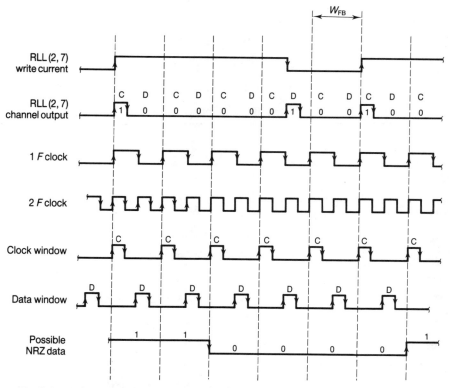

Fig. 7.1. Write waveform for the RLL (2, 7) code as used in digital magnetic recording.

following properties:

(A) A high density ratio (DR) that maximizes the total capacity of the disk drive.
(B) An acceptably low frequency ratio (FR) that facilitates the circuit design, low-pass filter, differentiator, signal-to-noise ratio, etc.

A cursory view of table 7.2. shows that the code RLL $(2, 7)$ is clearly superior to the others. This is the main reason why the RLL $(2, 7)$ is superseding MFM as the code of preference for modern disk drives.

However, if we add a third requirement:

(C) An acceptably high detection window w that facilitates the clock design,

the RLL $(2, 7)$ offers one of the lowest detection windows, namely 50% or the same as MFM. Therefore, for the particular case when the clock design is all-important, there are a number of other codes with window values higher than 50%.

The result is that we are left again with a decision that involves a compromise. However, due to their mathematical evaluation, it becomes possible to use actual figures of merit.

7.2. REPRESENTATION

It was stated in the previous chapter that, sometimes, the decision logic for the present bit takes into consideration the next bit and/or the previous bit. It is possible to expand this concept and consider many bits before and after, as shown in Fig. 7.2. by means of shift registers.

Regarding the bit under consideration, a decision is made based on the previous and future bits (which could be either data or clock). When this is done, the amount of possible codes becomes astronomical. However, from the point of view of magnetic transitions, some codes are more practical than others. In this regard, the decision logic is based on the following criteria:

(A) How many transition separations (minimum) can the system tolerate in terms of an acceptable pulse shift.
(B) How many transition separations (maximum) can the system tolerate in terms of an acceptable clock.

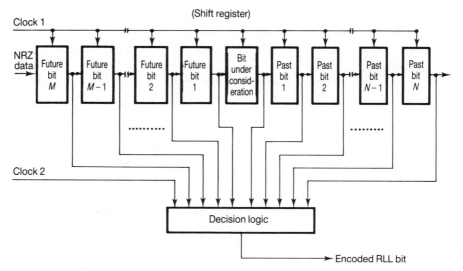

Fig. 7.2. Bit (either clock or data) encoded according to previous and past bits. *Conclusion*: The NRZ bit under consideration is encoded by considering N NRZ bits past and M NRZ bits future.

The requirements (A) and (B) are basically contradictory, so the final decision becomes a typical engineering compromise (in engineering, we do not have the luxury of believing that the world is perfect). In spite of these practical restrictions, the number of combinations would still be astronomical.

Let us restrict outselves to one particular case (the others become simple variations) and consider the parameters:

$$d = 2$$

$$k = 7$$

the Run-Length Limited code that results, namely, RLL $(2, 7)$ is defined as follows:

(A)   The transitions can be separated by no less than two.
(B)   The transitions can be separated by no more than seven.

However, with this meager information, it is still impossible to define accurately the "decision logic." The reason is simple: Even by fulfilling

Table 7.3.

ENDEC for the RLL (2, 7) Code Used by IBM Represented in Shift Register Form. (1) MSB = Most Significant Bit. Always First. (2) In the Section Entitled Coded Words, the Following Convention is Being Used: 0: No Transition, 1: Transition (Either Up-going or Down-going). (3) Always $n = 2m$. (4) Encoding Done During Write (m to n). (5) Decoding Done During Read (n to m).

Data words (m)				Coded words (n)							
S.R. 4	S.R. 3	S.R. 2	S.R. 1	Data transition 4	Clock transition 4	Data transition 3	Clock transition 3	Data transition 2	Clock transition 2	Data transition 1	Clock transition 1
1	0			0	1	0	0				
1	1			1	0	0	0				
0	0	0		1	0	0	1	0	0		
0	1	0		0	0	0	1	0	0		
0	1	1		0	0	1	0	0	0		
0	0	1	0	0	0	1	0	0	1	0	0
0	0	1	1	0	0	0	0	1	0	0	0
MSB				MSB							

127

the requirements above, it is still possible to obtain an astronomical number of combinations. This simple observation makes the code ambiguous up to this point.

To avoid ambiguity, out of the thousands of combinations possible, one particular encode/decode (ENDEC) table is illustrated that was arrived at with the purpose of minimizing P. Table 7.3. shows the ENDEC table used by IBM (obviously, not the only one possible) in shift register form. Figure 7.3. is a block diagram of the functions to be performed.

As stated before, it is possible to make a pictorial representation of a code, as shown in Fig. 6.1. This task is greatly simplified by effecting a time-reversal of the shift register table. Consequently, to make a waveform representation of the RLL (2, 7) code, let us timereverse Table 7.3. The end result is illustrated as Table 7.4.

This table can be used to obtain a pictorial representation of our particular RLL (2, 7) code. The seven data words with ther corresponding waveforms are shown in Fig. 7.4. Now it becomes possible to concatenate a large number of data bits, as done before, and obtain the corresponding RLL (2, 7) write waveform. This is illustrated in Fig. 7.5., which consists of an arbitrary string of ones and zeros with the corresponding write waveform that conforms to our particular ENDEC table.

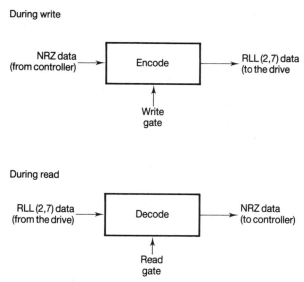

Fig. 7.3. Block diagram of the encode/decode (ENDEC) operations.

Table 7.4.

ENDEC for the RLL (2, 7) Code Used by IBM with a Time Reversal of the Shift Register Representation. (1) MSB = Most Significant Bit. Always First. (2) In the Section Entitled Coded Words, the Following Convention is Being Used: 0: No Transition, 1: Transition (Either Up-going or Down-going). (3) Always $n = 2m$. (4) Encoding Done During Write (m to n). (5) Decoding Done During Read (n to m).

Data words (m)				Coded words (n)							
S.R. 1	S.R. 2	S.R. 3	S.R. 4	Clock transition 1	Data transition 1	Clock transition 2	Data transition 2	Clock transition 3	Data transition 3	Clock transition 4	Data transition 4
		0	1					0	0	1	0
		1	1					0	0	0	1
	0	0	0			0	0	1	0	0	0
	0	1	0			0	0	1	0	0	1
	1	1	0			0	0	0	1	0	0
0	1	0	0	0	0	1	0	0	1	0	0
1	1	0	0	0	0	0	1	0	0	0	0

MSB (Data words)   MSB (Coded words)

129

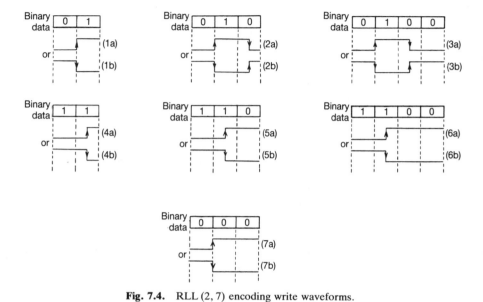

Fig. 7.4. RLL (2, 7) encoding write waveforms.

7.3. READ/WRITE WAVEFORMS

For purposes of comparison between the codes MFM and RLL (2, 7), let us show an example of their respective read and write waveforms. This is done in Fig. 7.6. where, for purposes of simplification, we are assuming 100% magnetic recording readback (zero bit shift). In either case, the worst case is shown, namely, two consecutive "ones." However, in the RLL (2, 7) case, it seems rather extravagant to allow pulses to be separated by such a large distance when in reality the magnetic recording system allows them to be "compacted" until the two closest "ones" will be next to each other without any wasted space between them.

Consequently, in the RLL (2, 7) case, the write process will include the encoding as well and will consist of the following operations:

(A) Receive the NRZ binary data (strings of "ones" and "zeros") that are going to be stored in the disk drive.
(B) Encode that data into the RLL (2, 7) form given by either Table 7.3. or Table 7.4.
(C) Based on that encode data, compact the data by means of the write clock and by using the appropriate compact ratio $DR = 1.5$.
(D) Generate from the above the corresponding write waveform.

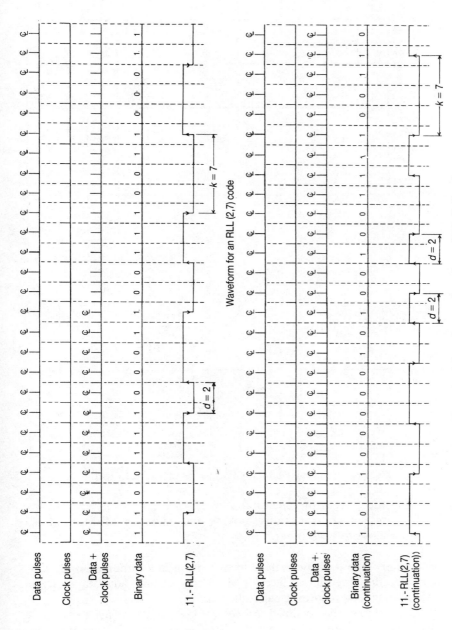

Fig. 7.5. Example of the write waveform for an RLL (2, 7) code.

131

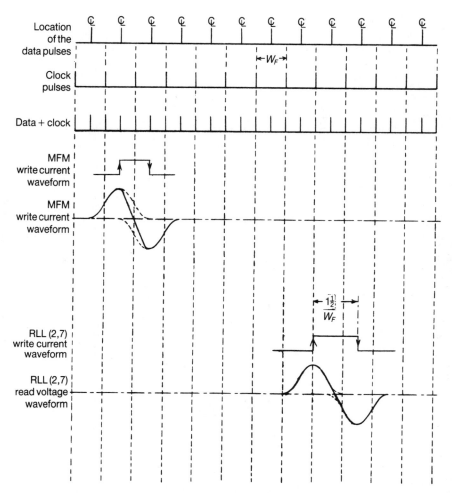

Fig. 7.6. Write and read waveforms for MFM and RLL (2, 7).

(E) Using the waveform generated above, write on the disk surface in two consecutive transitions looking the same as MFM, that is, rather close to each other.

In summary, the transitions have been written in a rather compact way in the drive, looking almost like MFM, simply to take advantage of the disk drive magnetic recording capabilities.

However, during the readback process, it became necessary to "undo" the write process and create back the NRZ waveform. This means that, in reality, the read process is nothing more than the opposite of compaction namely, "expansion," but not arbitrarily. That is, the expansion must be done while recreating exactly the same data that was received originally.

Consequently, the reading process will include the operation of decoding and will consist of the following operations:

(A)   Read from the disk by means of the read channel.
(B)   Expand the data. This is done by means of a shift register.
(C)   Decode the contents of the shift register by means of the decision (or combinatorial) logic. This operation gives a string of ones and zeros that should be exactly like the data that was originally intended to be stored.

Therefore, by this method of compaction and expansion, done with shift registers plus encode and decode logic, it becomes possible to make the outside world think that the disk drive can store more information (or that it has more capacity than it really has).

The price to pay for this higher performance is:

(A)   It becomes necessary to add extra hardware, such as shift registers, encode logic, decode logic, etc.
(B)   The requirements on the clocking system are very stringent. This is due to the fact that the clock has been multiplied by a factor DR, thus aggravating the clock imperfections—window margins, jitter, vibrations, distortion, etc. Consequently, a very good clock becomes mandatory.
(C)   Since the disk clock has been multiplied, the data comes out multiplied also, thus, the data rate is higher than the flux rate. See Eq. (3.7).

7.4. ENCODE

The overall design is based on Patent No. 4,115,768 (Eggenberger and Hodges, 1978). The encoding can be done three different but equivalent ways:

(A)   Software,
(B)   Firmware, and
(C)   Hardware.

The implementation by means of the first two methods above can be

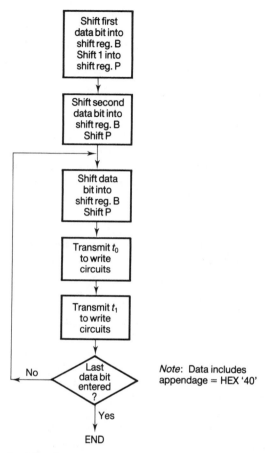

Fig. 7.7. Encoding procedure in flow diagram form.

facilitated by the flow chart shown in Fig. 7.7. An example is shown in Fig. 7.8. The encoder logic is shown in Fig. 7.9., and Fig. 7.10. shows an actual implementation using shift registers that implement the code of Table 7.3.

Notes.

(A) Encoding is done bit-by-bit rather than according to the varying length words shown in Table 7.3. ($r = 3$, that is, we have words of 2-bit length, 3-bit length and 4-bit length). The shift registers B and P are shifted once for each bit encoded.

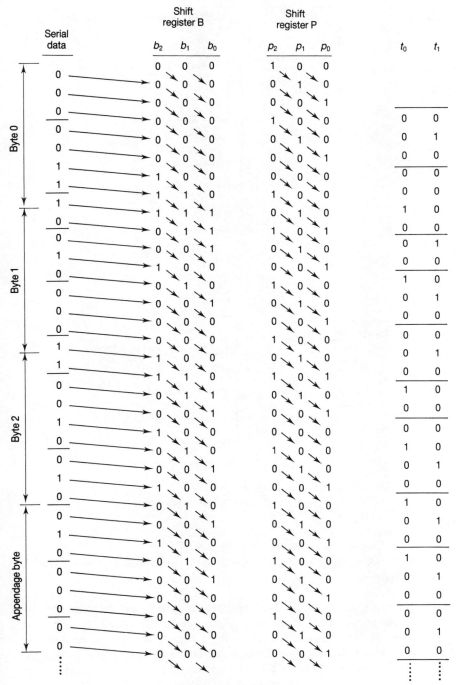

Fig. 7.8. Encoding example.

135

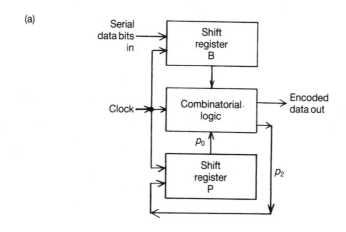

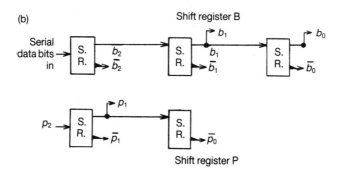

(c)

$$p_2 = (b_0 \cdot b_1 \cdot \overline{p_1}) + (\overline{b_0} \cdot b_1) + (\overline{b_0} \cdot \overline{b_1} \cdot \overline{b_2} \cdot \overline{p_0} \cdot \overline{p_1})$$
$$t_0 = (b_0 \cdot b_1 \cdot \overline{p_0}) + (\overline{b_0} \cdot b_1 \cdot \overline{b_2} \cdot \overline{p_0}) + [b_1 \cdot \overline{p_0} \cdot (b_0 + \overline{b_0} \cdot \overline{b_2})]$$
$$t_1 = b_1 \cdot p_1$$

Encoded data out $= \overline{(\text{clock})} \cdot t_0 + (\text{clock}) \cdot t_1$

Fig. 7.9. Encoder logic (during write, m to n). (a) Block diagram. (b) Shift registers. (c) Boolean expressions.

(B) The output P_2 and shift register P are used to account for boundaries since it is not always possible to determine the boundary between data words by observing a fixed number of consecutive bits. The value of P_2 is 1 when the last bit entered into B was the end of a word. Thus, $P_0(P_1)$ is 1 when the last bit of a word is in b_0 (b_1).

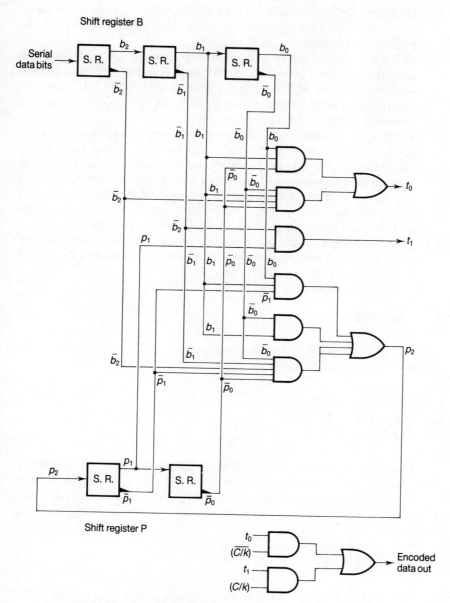

Note: Illustrated for the case of $n = 4$ (four bands)

Fig. 7.10. Serial encoder.

(C)   On entering the first data bit into B, the value of P_2 is made equal
       to 1, i.e., the first data bit of a sequence is defined to be the
       beginning of a word.
(D)   There is a two-bit delay associated with encoding. That is, the first
       three bits of data must be entered into B before the first pair of
       transition values is determined, and two dummy bits must be
       entered to enable determination of the last two pairs of transition
       values in the recorded sequence.
(E)   For proper encoding, it is necessary to ensure completion of a word
       at the end of the sequence of data bits. As noted later under
       "decoding", one dummy bit (beyond the desired data) must be
       entered and encoded, and, as noted above, two additional dummy
       bits must be entered to complete the encoding of the first (this is
       like "priming" the pump). Any bit pattern is acceptable, and
       additional dummy bits may be appended as seems desirable.
(F)   This has been a rather cursory view of the "encode" process. For
       more details of implementation, the reader is referred to the above
       mentioned Patent No. 4,115,768.

7.5. DECODE

Same as with "encode," the decoding process can be performed three
different but equivalent ways:

(A)   Software,
(B)   Firmware, and
(C)   Hardware.

   The implementation by means of the first two methods above can be
facilitated by the flow chart shown in Fig. 7.11. An example is given in
Fig. 7.12. The decoder logic is shown in Fig. 7.13. and Fig. 7.14. shows
an implementation using shift registers that implement the code of
Table 3.

Notes.

(A)   Same as was done with encoding, the decoding is done bit-by-bit
       rather than according to varying length words.
(B)   For each clock period a transition value (1 = transition, 0 = no

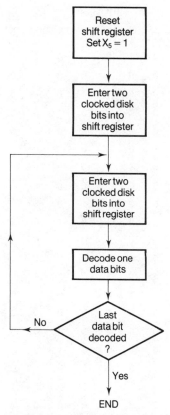

Fig. 7.11. Serial decoding procedure.

transition) is entered into the shift register $(X_0, X_1, X_2, \ldots, X_7)$. One data bit is decoded whenever an even number of transition values have been entered.

(C) There is a delay associated with decoding. The first four transition values must be entered into the shift register before the first data bit is decoded.

Since two transition values must be entered for each subsequent bit decoded, two properly encoded transition values must have been written after the pair corresponding to the last data bit.

(D) Again, this has been a rather cursory view of the "decode" process. For more details of implementation the reader is referred to the above mentioned Patent No. 4,115,768.

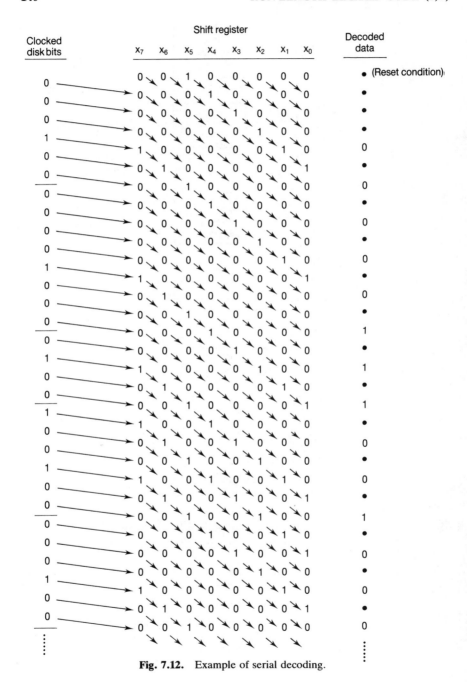

Fig. 7.12. Example of serial decoding.

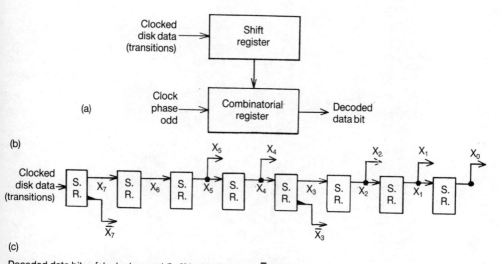

(c)

Decoded data bit = [clock phase odd] · [(X_2+X_4 · X_7)+(X_1 · $\bar{X}_3$ · X_5)+(X_0 · X_5)]

Fig. 7.13. Decoder logic (during read, *n* to *m*). (a) Block diagram. (b) Shift registers. (c) Boolean expressions.

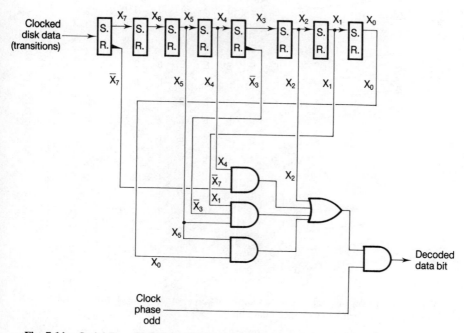

Fig. 7.14. Serial Decoder. Note: "Odd" clock implies an even number of bits entered.

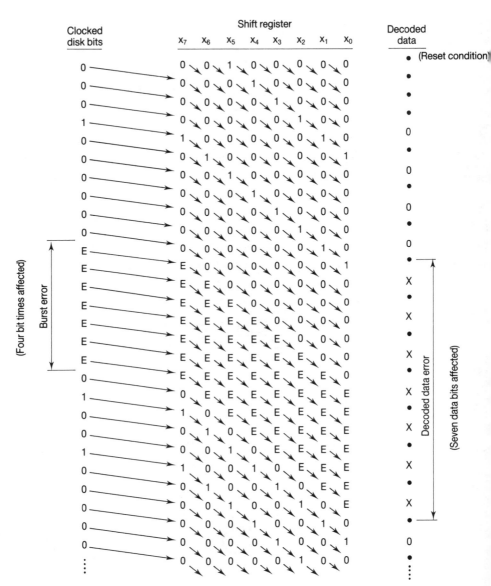

Fig. 7.15. Error propagation example. E denotes an error in a transition value. X denotes a possible error in a decoded data bit.

7.6. ERROR PROPAGATION AND CHECKING

A transition error can cause a number of bits of decoded data to be in error. With the shift register decoded, each transition affects four consecutive data bits. Thus, a data error due to an error in a transition value will be confined to four bits in the vicinity of the transition error, and a burst error will propagate on decoding for no more than three data bits beyond the location of the transitions in error. An example is shown in Fig. 7.15.

Encoder errors produce permanent errors in recorded data. This decoder is much more complex than those used in previous disk drives using MFM, NRZI, etc. Consequently, it is very desirable to check its operation.

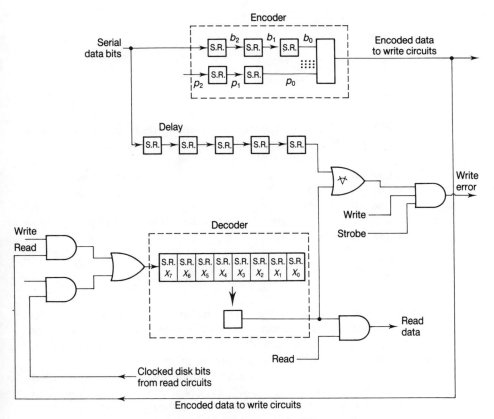

Fig. 7.16. Encoder error checking.

Table 7.5

Hexadecimal/Binary ENDEC.

Hexadecimal	Binary			
	2^3	2^2	2^1	2^0
0	0	0	0	0
1	0	0	0	1
2	0	0	1	0
3	0	0	1	1
4	0	1	0	0
5	0	1	0	1
6	0	1	1	0
7	0	1	1	1
8	1	0	0	0
9	1	0	0	1
A	1	0	1	0
B	1	0	1	1
C	1	1	0	0
D	1	1	0	1
E	1	1	1	0
F	1	1	1	1

This is done by encoding the incoming data, then decoding it and comparing it with the incoming data bits. The results of these complementary operations (encode plus decode) should be exactly equal to the original incoming bits (after the proper delay to assure synchronization). Any discrepancies are flagged as errors. This rather ingenious scheme is due to Dr. Paul Hodges (Hodges, 1980) and is thoroughly covered in detail in the article mentioned in the Bibliography. Figure 7.16 shows this scheme in block form.

7.7. TESTING

As with any coding scheme, there are several useful patterns that can be used to test the recording characteristics of the disk file. These patterns are chosen in such a way that they test some important parameter, such as bit shift, maximum frequency, minimum frequency, etc. Table 7.5. shows the ENDEC for the standard hexadecimal notation and is included here for reference purposes. However, Figs. 7.17. to 7.21. show several patterns which have been found useful while testing the R/W recording characteristics of the disk drive.

Table 1 (NRZ cell no. 1–28 / RLL(2,7) cell no. 1–56)

Hexadecimal	B				5				6				A				D				5				A			
NRZ cell no.	1	2	3	4	5	6	7	8	9	10	11	12	13	14	15	16	17	18	19	20	21	22	23	24	25	26	27	28
NRZ data	1	0	1	1	0	1	0	1	0	1	1	0	1	0	1	0	1	1	0	1	0	1	0	1	1	0	1	0

Word markers (top to bottom): 2-Bit word, 2-Bit word, 3-Bit word, 2-Bit word, 2-Bit word, 3-Bit word, 2-Bit word, 2-Bit word, 3-Bit word, 2-Bit word, 2-Bit word, 3-Bit word, 2-Bit word …

RLL(2,7) cell no. 1–56, RLL(2,7) encoding (each NRZ cell = 2 RLL cells).

Table 2 (NRZ cell no. 29–56 / RLL(2,7) cell no. 57–112)

Hexadecimal	B				B				B				5				6				5							
NRZ cell no.	29	30	31	32	33	34	35	36	37	38	39	40	41	42	43	44	45	46	47	48	49	50	51	52	53	54	55	56
NRZ data	1	0	1	1	1	0	1	1	0	1	0	1	0	1	1	0	1	0	1	1	0	1	0	1	1	0	1	1

RLL(2,7) cell no. 57–112. ← Boundary (four "0" at boundary — worse case).

Table 3 (NRZ cell no. 57–84 / RLL(2,7) cell no. 113–168)

Hexadecimal	A				A				D				A				5				B				D			
NRZ cell no.	57	58	59	60	61	62	63	64	65	66	67	68	69	70	71	72	73	74	75	76	77	78	79	80	81	82	83	84
NRZ data	1	0	1	0	1	0	1	1	1	0	1	1	0	1	0	1	0	1	1	0	1	0	1	0	1	1	0	1

RLL(2,7) cell no. 113–168. ← Boundary

Fig. 7.17. Example of MFM and RLL (2, 7) patterns for worst case shift testing. *Notes:* (1) HEX. B56A D5AB (8 × 4 = 32 Bits) produces an RLL (2, 7) pattern equivalent to the MFM 6DB6, that is, worse case bit shift pattern, except at the boundary where there are four (4) '0' instead of three (3). (2) Going from top to bottom produces NRZ to RLL (2, 7) encoding (3) Going from bottom to top produces RLL (2, 7) to NRZ decoding.

Fig. 7.18. Example of an RLL (2, 7) pattern that produces ascending and descending "ones". *Notes:* (1) HEX. F819 801A (8 × 4 = 32 bits) produces an ascending and descending series of "1" in RLL (2, 7). (2) Going from top to bottom produces NRZ to RLL (2, 7) encoding. (3) Going from bottom to top produces RLL (2, 7) to NRZ decoding.

Table 1 (top):

Hexadecimal	B				0				9				9				1				8				2			
NRZ cell no.	1	2	3	4	5	6	7	8	9	10	11	12	13	14	15	16	17	18	19	20	21	22	23	24	25	26	27	28
NRZ data	1	0	1	1	0	0	0	0	1	0	0	1	1	0	0	1	0	0	0	1	1	0	0	0	0	0	1	0
	2-Bit word				2-Bit word				3-Bit word				3-Bit word				3-Bit word				4-Bit word				4-Bit word			

RLL(2,7) cell no.	1	2	3	4	5	6	7	8	9	10	11	12	13	14	15	16	17	18	19	20	21	22	23	24	25	26	27	28	29	30	31	32	33	34	35	36	37	38	39	40	41	42	43	44	45	46	47	48	49	50	51	52	53	54	55	56
RLL(2,7) encoding	0	1	0	0	1	0	0	0	0	0	0	1	0	0	1	0	0	1	0	0	0	0	1	0	0	0	0	0	1	0	0	1	0	0	0	0	0	1	0	0	0	0	0	1	0	0	0	0	1	0	0	1	0	0		

Table 2 (middle):

Hexadecimal	3				B				0				9				9				1				8		
NRZ cell no.	29	30	31	32	33	34	35	36	37	38	39	40	41	42	43	44	45	46	47	48	49	50	51	52	53	54	55	56
NRZ data																											
	4-Bit word				2-Bit word		2-Bit word		3-Bit word			3-Bit word			3-Bit word			4-Bit word			4-Bit word			3-Bit word			

RLL(2,7) cell no.	57	58	59	60	61	62	63	64	65	66	67	68	69	70	71	72	73	74	75	76	77	78	79	80	81	82	83	84	85	86	87	88	89	90	91	92	93	94	95	96	97	98	99	100	101	102	103	104	105	106	107	108	109	110	111	112
RLL(2,7) encoding	0	0	0	0	1	0	0	0	0	1	0	0	1	0	0	0	0	0	0	1	0	0	1	0	0	1	0	0	0	0	1	0	0	0	0	0	1	0	0	1	0	0	0	0	0	0	1	0	0	0	0	0	1	0	0	

↑ — Boundary

Table 3 (bottom):

Hexadecimal	2				3				B				0				9				9				1		
NRZ cell no.	57	58	59	60	61	62	63	64	65	66	67	68	69	70	71	72	73	74	75	76	77	78	79	80	81	82	83	84
NRZ data	0	0	1	0	0	0	1	1	1	0	1	1	0	0	0	0	1	0	0	1	1	0	0	1	0	0	0	1
	4-Bit word				4-Bit word				2-Bit word		2-Bit word		3-Bit word			3-Bit word			3-Bit word			4-Bit word			3-Bit word		

RLL(2,7) cell no.	113	114	115	116	117	118	119	120	121	122	123	124	125	126	127	128	129	130	131	132	133	134	135	136	137	138	139	140	141	142	143	144	145	146	147	148	149	150	151	152	153	154	155	156	157	158	159	160	161	162	163	164	165	166	167	168
RLL(2,7) encoding	0	0	0	0	0	1	0	0	0	0	0	0	1	0	0	0	0	1	0	0	1	0	0	0	0	0	0	1	0	0	1	0	0	1	0	0	0	0	1	0	0	0	0	0	1	0	0	1	0	0	0	0	0	0	1	0

↑ — Boundary

Fig. 7.19. Example of an RLL (2, 7) pattern that tests the complete ENDEC table. *Notes:* (1) The HEX. B099 1823 ($8 \times 4 = 32$ Bits) checks out the whole RLL (2, 7) ENDEC table. (2) Going from top to bottom produces NRZ to RLL (2, 7) encoding. (3) Going from bottom to top produces RLL (2, 7) to NRZ decoding.

Table 1 — NRZ cells 1–28 / RLL(2,7) cells 1–56

Hexadecimal	9				2				4				9				...				

NRZ cell no.	1	2	3	4	5	6	7	8	9	10	11	12	13	14	15	16	17	18	19	20	21	22	23	24	25	26	27	28
NRZ data	1	0	0	1	0	0	1	0	0	1	0	0	1	0	0	1	0	0	1	0	0	1	0	0	1	0	0	1

RLL(2,7) words: 2-Bit word, then 3-Bit words repeated.

RLL(2,7) cell no.	1	2	3	4	5	6	7	8	9	10	11	12	13	14	15	16	...	56
RLL(2,7) encoding	0	1	0	0	1	0	0	1	0	0	1	0	0	1	0	0	...	1

— Boundary —

Table 2 — NRZ cells 29–56 / RLL(2,7) cells 57–112

Hexadecimal	2				9				4				2	...

NRZ cell no.	29	30	31	32	33	34	35	36	37	38	39	40	41	42	43	44	45	46	47	48	49	50	51	52	53	54	55	56
NRZ data	0	0	1	0	0	1	0	0	1	0	0	1	0	0	1	0	0	1	0	0	1	0	0	1	0	0	1	0

RLL(2,7) words: 3-Bit words.

RLL(2,7) cell no.	57	58	59	60	61	62	63	64	65	66	67	68	...	112
RLL(2,7) encoding	0	0	1	0	0	1	0	0	1	0	0	1	...	0

— Boundary —

Table 3 — NRZ cells 57–84 / RLL(2,7) cells 113–168

Hexadecimal	9				2				4				2	...

NRZ cell no.	57	58	59	60	61	62	63	64	65	66	67	68	69	70	71	72	73	74	75	76	77	78	79	80	81	82	83	84
NRZ data	1	0	0	1	0	0	1	0	0	1	0	0	1	0	0	1	0	0	1	0	0	1	0	0	1	0	0	0

RLL(2,7) words: 3-Bit words with a 2-Bit word.

RLL(2,7) cell no.	113	114	115	116	117	118	119	120	...	165	166	167	168
RLL(2,7) encoding	0	1	0	0	1	0	0	1	...	0	0	0	0

Fig. 7.20. Example of an RLL (2, 7) pattern that tests for the *highest* possible frequency. *Notes:* (1) HEX. 9249 2492 (8 × 4 = 32 Bits) produces the *highest* possible frequency for RLL (2, 7), namely, a "1" surrounded by two (2) "0", except at the boundary, where there are three (3) "0" instead of only two. (2) Going from top to bottom produces NRZ to RLL (2, 7) encoding. (3) Going from bottom to top produces RLL (2, 7) to NRZ decoding.

Hexadecimal	3								3								3								3								3								3								3							
NRZ cell no.	1		2		3		4		5		6		7		8		9		10		11		12		13		14		15		16		17		18		19		20		21		22		23		24		25		26		27		28	
NRZ data	0		0		1		1		0		0		1		1		0		0		1		1		0		0		1		1		0		0		1		1		0		0		1		1		0		0		1		1	
	4-Bit word								4-Bit word								4-Bit word								4-Bit word								4-Bit word								4-Bit word								4-Bit word							
RLL(2,7) cell no.	1	2	3	4	5	6	7	8	9	10	11	12	13	14	15	16	17	18	19	20	21	22	23	24	25	26	27	28	29	30	31	32	33	34	35	36	37	38	39	40	41	42	43	44	45	46	47	48	49	50	51	52	53	54	55	56
RLL(2,7) encoding	0	0	0	0	1	0	0	0	0	0	0	0	1	0	0	0	0	0	0	0	1	0	0	0	0	0	0	0	1	0	0	0	0	0	0	0	1	0	0	0	0	0	0	0	1	0	0	0	0	0	0	0	1	0	0	0

Hexadecimal	3								3								3								3								3								3								3							
NRZ cell no.	29		30		31		32		33		34		35		36		37		38		39		40		41		42		43		44		45		46		47		48		49		50		51		52		53		54		55		56	
NRZ data	0		0		1		1		0		0		1		1		0		0		1		1		0		0		1		1		0		0		1		1		0		0		1		1		0		0		1		1	
	4-Bit word								4-Bit word								4-Bit word								4-Bit word								4-Bit word								4-Bit word								4-Bit word							
RLL(2,7) cell no.	57	58	59	60	61	62	63	64	65	66	67	68	69	70	71	72	73	74	75	76	77	78	79	80	81	82	83	84	85	86	87	88	89	90	91	92	93	94	95	96	97	98	99	100	101	102	103	104	105	106	107	108	109	110	111	112
RLL(2,7) encoding	0	0	0	0	1	0	0	0	0	0	0	0	1	0	0	0	0	0	0	0	1	0	0	0	0	0	0	0	1	0	0	0	0	0	0	0	1	0	0	0	0	0	0	0	1	0	0	0	0	0	0	0	1	0	0	0

Hexadecimal	3								3								3								3								3								3								3							
NRZ cell no.	57		58		59		60		61		62		63		64		65		66		67		68		69		70		71		72		73		74		75		76		77		78		79		80		81		82		83		84	
NRZ data	0		0		1		1		0		0		1		1		0		0		1		1		0		0		1		1		0		0		1		1		0		0		1		1		0		0		1		1	
	4-Bit word								4-Bit word								4-Bit word								4-Bit word								4-Bit word								4-Bit word								4-Bit word							
RLL(2,7) cell no.	113	114	115	116	117	118	119	120	121	122	123	124	125	126	127	128	129	130	131	132	133	134	135	136	137	138	139	140	141	142	143	144	145	146	147	148	149	150	151	152	153	154	155	156	157	158	159	160	161	162	163	164	165	166	167	168
RLL(2,7) encoding	0	0	0	0	1	0	0	0	0	0	0	0	1	0	0	0	0	0	0	0	1	0	0	0	0	0	0	0	1	0	0	0	0	0	0	0	1	0	0	0	0	0	0	0	1	0	0	0	0	0	0	0	1	0	0	0

Fig. 7.21. Example of an RLL (2, 7) pattern which tests for the lowest possible frequency. *Notes:* (1) The Hex. 3, as shown above, produces the *lowest* possible frequency of "1" after RLL (2, 7) encoding. (2) Going from top to bottom produces NRZ to RLL (2, 7) encoding. (3) Going from bottom to top produces RLL (2, 7) to NRZ decoding.

7.8. SUMMARY

The implementation of the RLL (2, 7) code has been detailed as an example of the requirements to be fulfilled for whichever code is used. The encode/decode operations have been detailed in hardware and software form, along with several patterns useful for testing operations. As with all codes, if the implementation is done by specialized circuitry (hardware), the encode/decode becomes transparent to the user. If this hardware is absent, then it must be done by computer programming, using the flow diagrams shown above for this particular code. This is a very dynamic section of DASDs and consequently, it is expected that in the near future many new codes will be proposed and implemented.

8

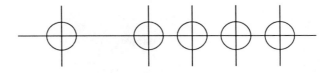

Read/Write Channel

DASDs are useful because they can store (the write process) great quantities of data for a long period of time. Later on, that data can be retrieved non-destructively (the read process) at random. The write data (plus corresponding clock) is generated by a computer, and the electronic circuitry used to convey and deposit that information on the recording surface is named the "write channel." When retrieving that data and clock (under computer command), the magnetically recorded transitions are first detected by the read head and thereafter sent to a series of circuits named the "read channel" until the data and clock are in such form that they can be used by the computer. This chapter is an introduction to the circuit functions and components that are needed to perform these operations in a reliable fashion. The actual implementation of the circuits fall into the realm of circuit design that is beyond the scope of this book. Consequently, only the most critical components will be detailed.

8.0. READ/WRITE CHANNEL

The Read/Write (R/W) channel is the electronic circuitry that stands between the controller and the R/W head. At the beginning, this circuitry was implemented by means of vacuum tubes. Thereafter, transistors and later on, hybrid circuits were used. Today, they are implemented by means of integrated circuits (IC) with some attached passive components (resistors, capacitors, and/or inductors). However, the general block diagram has not changed much, as shown in Fig. 8.1. It is expected that, regardless of the technology being used for data recording, for the near future, this situation will not change either.

8.1. WRITE

This is a relatively easy function to implement, due to the nature of the magnetic recording phenomena being used:

(a)   Magnetic surface saturation. This is done very simply by means of

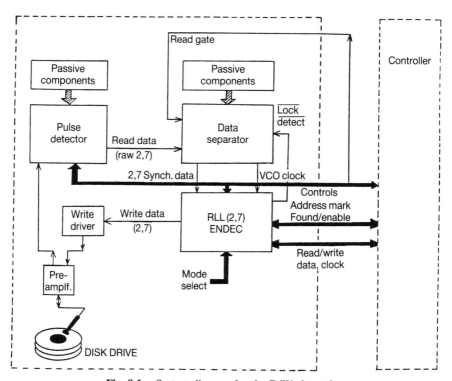

Fig. 8.1.   System diagram for the R/W channel.

an appropriate current that is generated by a current source. Thereafter, this current is sent to the selected R/W head by a write driver.

(b) Magnetic transition. On a magnetic recording surface the data is recorded by means of transitions, or changes of directions in the little magnets being created (north to south or south to north). These transitions are created very easily by means of a flip-flop (usually a toggle flip-flop) which changes state according to the code being used (see Chapters 6 and 7).

Usually, these components occupy so little space (in IC form) that several of them are packaged together, as seen in Fig. 8.2., and located as close as possible to the R/W heads. Usually, immediately next to the head flexures.

Since there are many heads and the data to be written arrives in a single wire, this data must be steered to the proper destination. This implies that some form of de/multiplexing must be used. At one time this function was implemented by means of a diode matrix. Today, this is no longer necessary, as shown in Fig. 8.2.

8.2. READ

The recorded signals being detected by the read head are extremely small, usually in the order of five millivolts. Since TTL circuits require at least five volts, it is necessary to effect a signal amplification of an approximate magnitude of 1,000. At the same time, these small signals arrive accompanied by a tremendous amount of noise, and after several stages of amplification (plus manipulations), the signal-to-noise ratio is rather poor. Consequently, the design of the read channel is very critical. The read channel consists of the following components:

(a)   Pre-amplifiers and multiplexer
(b)   VGA (variable gain amplifier)
(c)   AGC (automatic gain control)
(d)   Low-pass filter
(e)   Differentiator
(f)   Pulse discriminator
(g)   Data separator
(h)   Interface

Sometimes it also includes other helpful functions such as equalization, DC restoration, etc.

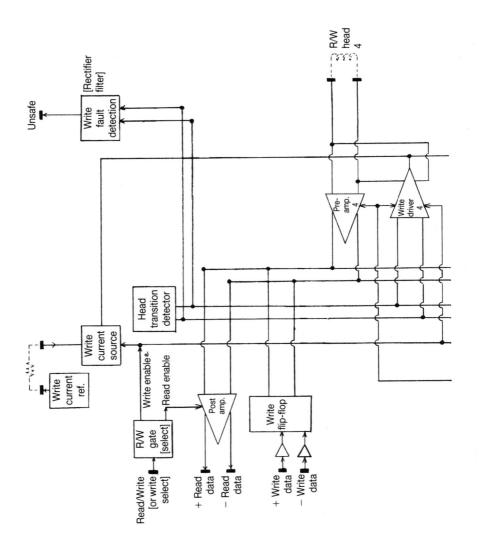

154

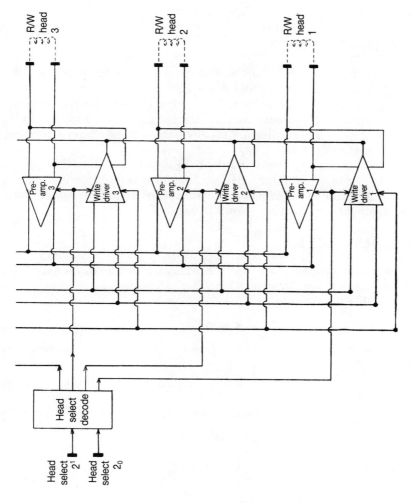

Fig. 8.2. Four pre–amplifiers and write drivers packaged in a single integrated circuit. Reproduced with permission from Silicon Systems, Inc.

155

Most of those components fall within the realm of circuit design and, since they are rather straightforward, they will not be mentioned anymore. However, there are some components that offer unique details that must be addressed here. Therefore, some of them will be treated in separate sections.

8.3. PRE-AMPLIFIER

The pre-amplifier is nothing more than a standard differential amplifier that is located as close as possible to the R/W head. Also, each head has its own pre-amplifier (and its own write driver). Since the requirements for write drivers and pre-amplifiers are the same (one for each head plus close proximity to them), they are usually packaged together in the same IC, as shown in Fig. 8.2.

Since there are many heads and the read data leaves this IC by means of a single pair of wires, there must be some form of multiplexing. At one time, this function was implemented by means of a diode matrix. Today, this is no longer necessary, as shown in Fig. 8.2.

8.4. EQUALIZER

This topic was addressed in Chapter 2. Many implementations have been suggested and/or implemented in the past. However, for a compression ratio K within the range

$$1 < K \leqslant 1.5, \tag{8.1}$$

the poles and zeros configuration shown in Fig. 8.3. has proven very successful. We are given

$$\text{Zeros:} \quad +a \pm jb \tag{8.2}$$

$$\text{Poles:} \quad -b \pm ja \tag{8.3}$$

where

$$a = \frac{1}{\phi}(1.27122988) \tag{8.4}$$

$$b = \frac{1}{\phi}(0.340625032) \tag{8.5}$$

and

$$\phi = \frac{\sqrt{K^2 - 1}}{2K} \tag{8.6}$$

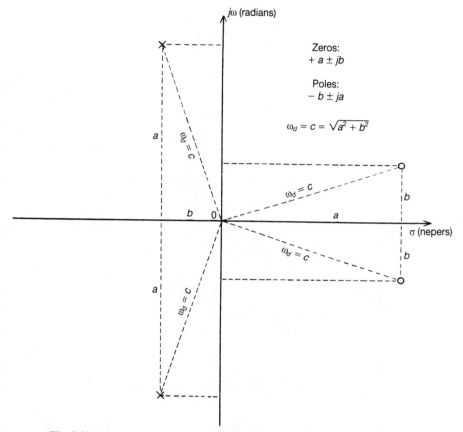

Fig. 8.3. Poles and zeros of an equalizer with a compression ratio $K \leqslant 1.5$.

Example: Let us consider a compression ratio

$$K = 1.5 \qquad (8.7)$$

Substituting (8.7) in (8.6)

$$\phi = \frac{\sqrt{1.25}}{3} = 0.372677996. \qquad (8.8)$$

substituting (8.8) in (8.4) and (8.5)

$$a = \frac{1.27122988}{0.372677996} = 3.41106771 \quad \text{and} \qquad (8.9)$$

$$b = \frac{0.340625032}{0.372677996} = 0.913992872, \qquad (8.10)$$

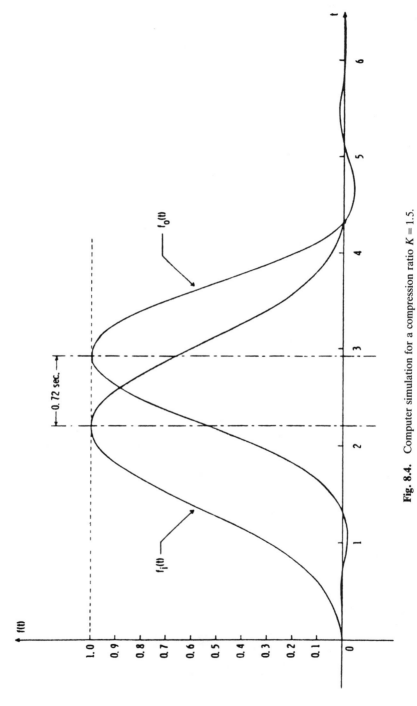

Fig. 8.4. Computer simulation for a compression ratio $K = 1.5$.

which gives the transfer function

$$H(S) = \frac{[(s - 3.41106771)^2 + (0.913992872)^2]}{[(s + 0.91399287)^2 + (3.41106771)^2]}.$$

That is

$$H(S) = \frac{s^2 - 6.82213542s + 12.47076589}{s^2 + 1.82798574s + 12.47076589}. \tag{8.11}$$

The Gaussian approximation $f_i(t)$ normalized to have a base width of

$$W_i = \sqrt{2}\,\pi = 4.4428829 \text{ secs} \tag{8.12}$$

and convolved with the inverse Laplace transform of (8.11) gives an output pulse $f_o(t)$ with a base width of

$$W_o = \frac{\sqrt{2}\,\pi}{1.5} = \frac{4.4428829}{1.5} = 2.961922 \text{ secs}. \tag{8.13}$$

The results obtained from the computer are shown in Fig. 8.4. The maximum value of $f_o(t)$ has been normalized to 1 and the delay from input to output is 0.72 secs.

To obtain the frequency response, let us divide (8.11) by the reciprocal of K and let us make $s = j\omega$, we obtain

$$H(j\omega) = \left(\frac{1}{1.5}\right) \frac{(12.47076589 - \omega^2) + j(-6.2213542\omega)}{(12.47076589 - \omega^2) + j(1.82798574\omega)}.$$

The magnitude function is

$$|H(j\omega)| = \left(\frac{1}{1.5}\right) \sqrt{\frac{(12.47076589 - \omega^2)^2 + (6.82213542\omega)^2}{(12.47076589 - \omega^2)^2 + (1.82798574\omega)^2}}. \tag{8.14}$$

Evaluating (8.14) for the interval

$$0 \leqslant \omega \leqslant 8 \text{ radians/sec.}$$

We obtain Fig. 8.5. This approximation seems to hold only up to ω_d for which Max $|H(jw)|$ occurs. This maximum happens when

$$12.47076589 - \omega^2 = 0$$

which gives

$$\omega_d = \sqrt{12.47076589} = 3.531397 \text{ radians/sec} \tag{8.15}$$

Substituting (8.15) in (8.14) we obtain

$$\text{Max } |H(j\omega)| = \left(\frac{1}{1.5}\right) \frac{6.82213542}{1.82798574} = 2.488034 \tag{8.16}$$

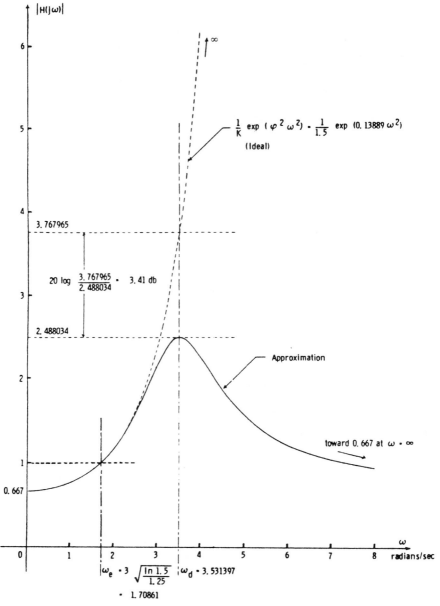

Fig. 8.5. Magnitude versus frequency for a compression ratio $K = 1.5$.

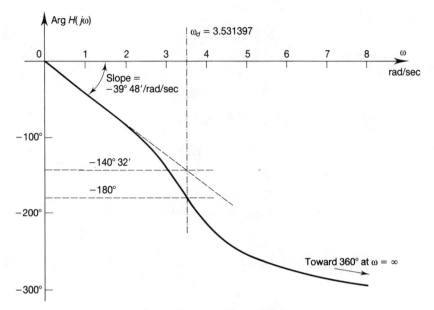

Fig. 8.6. Phase versus frequency for a compression ratio $K = 1.5$.

The phase is given by

$$\text{Arg } H(j\omega) = \arctan \frac{-6.82213542\omega}{12.47076589 - \omega^2} - \arctan \frac{1.82798574\omega}{12.47076589 - \omega^2} \quad (8.17)$$

Evaluating (8.17) for the interval

$$0 \leq \omega \leq 8 \text{ radians/sec.}$$

We obtain Fig. 8.6. It is observed that up to the frequency ω_d, the phase can be considered linear (no pulse distorsion) to an acceptable degree of accuracy.

For passive network synthesis, let us consider a constant R symmetrical lattice (not the only possibility). For this particular case, the series-arm impedance is given by

$$Z_a = \frac{1 - H(s)}{1 + H(s)}, \quad (8.18)$$

and the cross-arm impedance by

$$Z_b = \frac{1}{Z_a} \quad (8.19)$$

with the condition

$$|H(j\omega)| \leqslant 1 \quad \text{for} \quad 0 \leqslant \omega \leqslant \infty \qquad (8.20)$$

To fulfill requirement (8.20) it becomes necessary to multiply $H(s)$ by

$$A \leqslant \frac{1}{\text{Max } |H(j\omega)|}. \qquad (8.21)$$

Substituting (8.16) in (8.21), let us choose

$$A = \frac{1}{2.488034} \qquad (8.22)$$

which gives an attenuation of

$$20 \log(2.488034) = 7.92 \text{ dB}.$$

Multiplying (8.11) by (8.22), and by the reciprical of K as done before for the magnitude function, we obtain

$$H(s) = \frac{0.267949202s^2 - 1.82798574s + 3.34153177}{s^2 + 1.82798574s + 12.47076589}. \qquad (8.23)$$

Substituting (8.23) in (8.18)

$$Z_a = \frac{0.732050798s^2 + 3.65597149s + 9.12923412}{1.267949202s^2 + 15.81229766}$$

$$= 0.57735026 + \frac{2.88337378\,s}{s^2 + 12.47076589}. \qquad (8.24)$$

The synthesis of Z_a is shown in Table 8.1.
   Substituting (8.24) in (8.19)

$$Z_b = 1 \bigg/ \left(0.57735026 + \frac{2.88337378s}{s^2 + 12.47076589}\right). \qquad (8.25)$$

The synthesis of Z_b is shown also in Table 8.1.
   Combining Z_a and Z_b it becomes possible to obtain the normalized, constant R, symmetrical lattice shown in Fig. 8.7.
   The component values shown in Fig. 8.7. can be unnormalized by using the formulas

$$R_{\text{act}} = RR_n, \qquad (8.26)$$

$$L_{\text{act}} = \frac{L_n}{\sqrt{2} \times \pi}\, W_i R, \quad \text{and} \qquad (8.27)$$

$$C_{\text{act}} = \frac{C_n}{\sqrt{2} \times \pi} \times \frac{W_i}{R}. \qquad (8.28)$$

Table 8.1.
Individual Impedances for the Passive Filter Realization.

Impedance	Transfer function	Configuration
Z_a	$0.57735026 + \dfrac{2.88337378\,s}{s^2 + 12.47076589}$	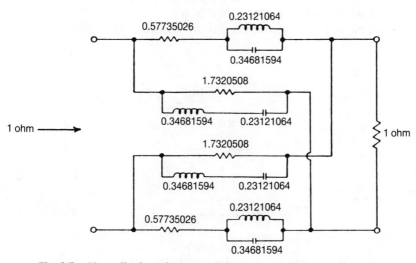
Z_b	$\dfrac{1}{0.57735026 + \dfrac{2.88337378\,s}{s^2 + 12.47076589}}$	

Fig. 8.7. Normalized passive network for a compression ratio $K = 1.5$.

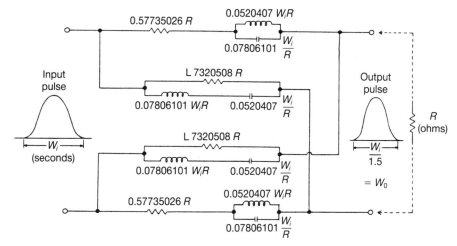

Fig. 8.8. Unnormalized passive network for a compression ratio $K = 1.5$.

Therefore, by using the normalized values R_n, L_n and C_n shown in Fig. 8.7. and applying Eqs. (8.26), (8.27) and (8.28), we obtain the unnormalized network shown in Fig. 8.8.

For purposes of illustration, let us consider the practical values $W_i = 2$ microseconds and $R = 2$ kilohms. Applying those values to Fig. 8.8, we obtain the actual network shown in Fig. 8.9.

For active network synthesis, let us consider the general network

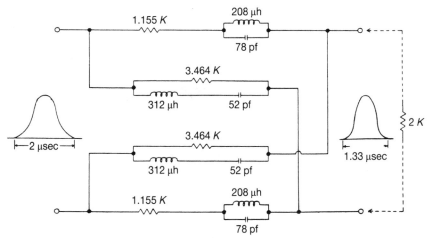

Fig. 8.9. Passive network for $K = 1.5$, $W_i = 2\,\mu\text{sec.}$, $R = 2$ Kilohms.

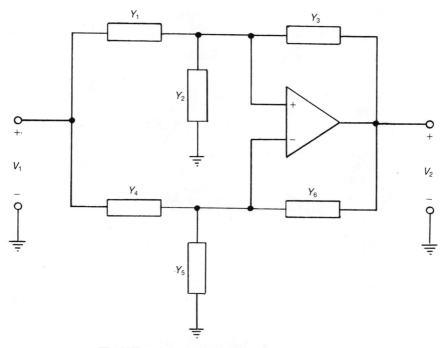

Fig. 8.10. A general active network configuration.

configuration shown in Fig. 8.10. (Huelsman, 1968, p. 203). The transfer
function for this network is given by the expression:

$$H(s) = \frac{V_2(s)}{V_1(s)} = \frac{Y_1(Y_4 + Y_5 + Y_6) - Y_4(Y_1 + Y_2 + Y_3)}{Y_6(Y_1 + Y_2 + Y_3) - Y_3(Y_4 + Y_5 + Y_6)}. \tag{8.29}$$

However, if we make

$$Y_1 + Y_2 + Y_3 = Y_4 + Y_5 + Y_6, \tag{8.30}$$

Eq. (8.29) reduces to

$$H(s) = \frac{V_2(s)}{V_1(s)} = \frac{Y_1 - Y_4}{Y_6 - Y_3}, \tag{8.31}$$

where Y = Admittance (reciprocal of impedance).
The poles and zeros shown in Fig. 8.3. give a transfer function

$$H(s) = \frac{(s-a)^2 + b^2}{(s+b)^2 + a^2} = \frac{s^2 - 2as + c^2}{s^2 + 2bs + c^2} \tag{8.32}$$

where

$$c^2 = a^2 + b^2. \tag{8.33}$$

In (8.32), let us divide numerator and denominator by $(s + c)$. Thus,

$$H(s) = \frac{\dfrac{s^2 + c^2}{s + c} - \dfrac{2as}{s + c}}{\dfrac{s^2 + c^2}{s + c} + \dfrac{2bs}{s + c}} = \frac{\left[(s + c) - \dfrac{2cs}{s + c}\right] - \left[\dfrac{2as}{s + c}\right]}{\left[(s + c) - \dfrac{2cs}{s + c}\right] + \left[\dfrac{2bs}{s + c}\right]}$$

$$= \frac{[s + c] - \left[\dfrac{2(c + a)s}{s + c}\right]}{[s + c] - \left[\dfrac{2(c - b)s}{s + c}\right]}. \tag{8.34}$$

Comparing (8.31) and (8.34), we can quickly identify

$$Y_1 = s + c, \tag{8.35}$$

$$Y_4 = \frac{2(c + a)s}{s + c}, \tag{8.36}$$

$$Y_6 = s + c \quad \text{and} \tag{8.37}$$

$$Y_3 = \frac{2(c - b)s}{s + c}. \tag{8.38}$$

To satisfy condition (8.30), let us choose

$$Y_5 = 0. \tag{8.39}$$

Substituting (8.35), (8.36), (8.37), (8.38) and (8.39) in condition (8.30) gives

$$(s + c) + Y_2 + \left[\frac{2(c - b)s}{s + c}\right] = \left[\frac{2(c + a)s}{s + c}\right] + 0 + (s + c),$$

which simplifies to

$$Y_2 = \frac{2(a + b)s}{s + c}. \tag{8.40}$$

The synthesis of the impedances Z_1, Z_2, Z_3, Z_4, Z_5, and Z_6 are shown in Table 8.2. Substituting these impedances in the network of Fig. 8.10., we obtain Fig. 8.11.

For a compression ratio $K = 1.5$, the a and b values were given by Eqs.

Table 8.2.
Individual Impedances for the Active Filter Realization.

Impedance	Transfer function	Configuration
Z_1	$\dfrac{1}{s+c}$	$\dfrac{1}{c}$ (resistor) and 1 (capacitor) in parallel
Z_2	$\dfrac{1}{2(a+b)} + \dfrac{1}{\dfrac{2(a+b)}{c}s}$	resistor $\dfrac{1}{2(a+b)}$, capacitor $\dfrac{2(a+b)}{c}$
Z_3	$\dfrac{1}{2(c-b)} + \dfrac{1}{\dfrac{2(c-b)}{c}s}$	resistor $\dfrac{1}{2(c-b)}$, capacitor $\dfrac{2(c-b)}{c}$
Z_4	$\dfrac{1}{2(c+a)} + \dfrac{1}{\dfrac{2(c+a)}{c}s}$	resistor $\dfrac{1}{2(c+a)}$, capacitor $\dfrac{2(c+a)}{c}$
Z_5	∞	Open circuit
Z_6	$\dfrac{1}{s+c}$	$\dfrac{1}{c}$ (resistor) and 1 (capacitor) in parallel

(8.9) and (8.10). Also, we can calculate

$$c = \omega_{\mathrm{d}} = \sqrt{a^2 + b^2} = \sqrt{(3.4110677)^2 + (9.91399287)^2} = \sqrt{12.47076589}$$

$$= 3.5313972. \tag{8.41}$$

substituting (8.9), (8.10), and (8.41) in the network shown in Fig. 8.11., we obtain the normalized network shown in Fig. 8.12.

It is possible to use the same unnormalized formulas employed before, namely, (8.26) and (8.28)—there is no need for inductors in an active network—thus giving the unnormalized network shown in Fig. 8.13.

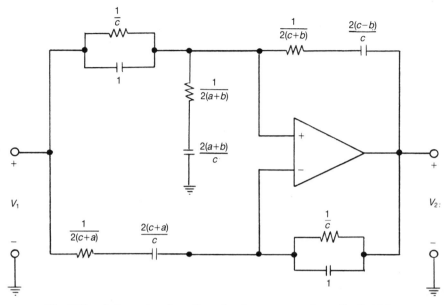

Fig. 8.11. Active network configuration for a compression ratio $K \leqslant 1.5$.

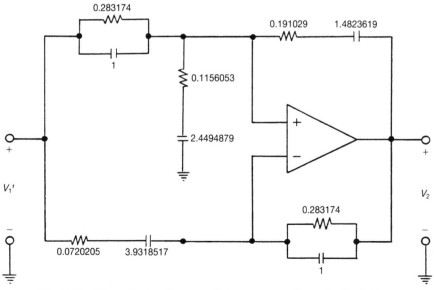

Fig. 8.12. Normalized active network for a compression ratio $K = 1.5$.

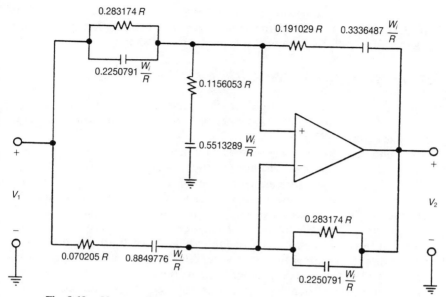

Fig. 8.13. Unnormalized active network for a compression ratio $K = 1.5$.

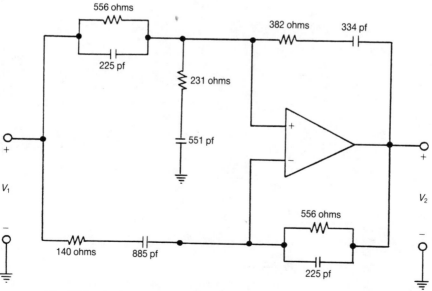

Fig. 8.14. Active network for $K = 1.5$, $W_i = 2\ \mu\text{sec.}$, $R = 2$ Kilohms.

For purposes of illustration let us consider the same values used before, namely, $W_i = 2$ microseconds, $R = 2$ kilohms. Applying those values to Fig. 8.13. we obtain the actual network shown in Fig. 8.14.

8.5. LOW-PASS FILTER

After the pre-amplifier/multiplexer chip, the rest of the read channel going all the way to the "data separator" constitutes the pulse detector circuit, as shown in Fig. 8.15.

As mentioned before, the signals obtained from the magnetic disks have a rather poor signal-to-noise ratio, usually exacerbated by the VGA–AGC loop and the equalizer (if used). Therefore, it becomes imperative to attenuate the high-frequency noise without distorting the read pulses. Those requirements imply the use of a low-pass filter with linear phase, which calls for another engineering compromise.

For that purpose, many types of filters have been tried and discarded. At the present time, it seems that acceptable results can be obtained by using a third-order, maximally flat delay (Bessel) filter. Although this type of filter has rather poor low-pass characteristics (not as good as Chebyshev, Legendre, Cauer, Butterworth, etc.), it offers linear phase at low-3dB cut-off frequencies, and this feature makes them very attractive for this particular application.

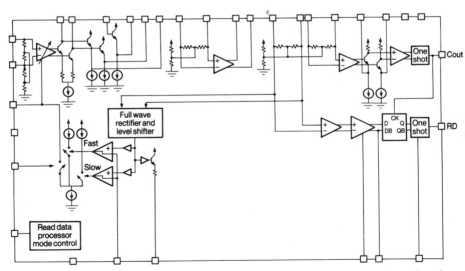

Fig. 8.15. Block Diagram of the pulse detector portion of a Read Channel. Reproduced with permission of Silicon Systems, Inc.

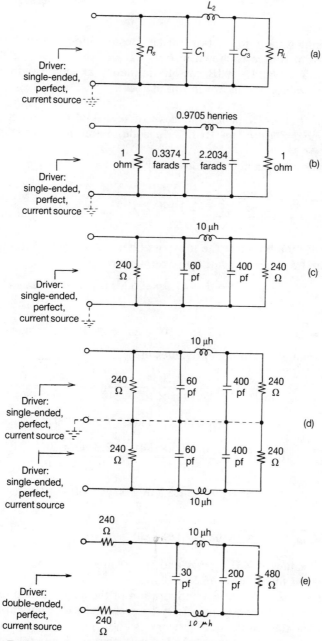

Fig. 8.16. Evolution of a single-ended low-pass filter to differential form. (a) From book. (b) Table values. (c) Unnormalized filter. (d) Two equal filters. (e) Result after combination.

The configuration for a third-order Bessel filter driven by a single-ended current source is shown in Fig. 8.16a (Zverev, 1967, p. 291). For $R_s = R_L = 1$ ohm and -3 dB cut-off frequency at $\omega_c = 1$ radian/sec, the normalized values are shown in Fig. 8.16b.

Example. Let us consider the case of a disk drive with a data rate $\Delta_B = 7.5$ megabits/sec, an RLL(2, 7) code. Filter load resistance $R_L = 240$ ohms.

The highest frequency F_H is given by

$$F_H = \frac{1}{2(d+1) \times \dfrac{m}{n} \times \dfrac{1}{\Delta_B}} = \frac{7.5 \times 10^6}{2(2+1) \times \frac{2}{4}} = \frac{7.5 \times 10^6}{3} = 2.5 \text{ MHz.} \quad (8.42)$$

The rule of thumb is: For the low-pass filter -3 dB cut-off frequency, use a factor of 1.5 times F_H. For our case we have

$$f_{-3 \text{ dB}} = 1.5 \times F_H = 1.5 \times 2.5 \text{ MHz} = 3.75 \text{ MHz}$$

which gives:

$$\omega_c = 2\pi f_{-3 \text{ dB}} = 2\pi \times 3.75 \times 10^6 = 23.561945 \times 10^6 \text{ radians/sec.}$$

The unnormalized component values are:

$$\text{CA:} \quad \frac{C_3}{\omega_c R_L} = \frac{2.2034}{23.561945 \times 10^6 \times 240} = \frac{2.3024 \times 10^{-12}}{0.0056548668}$$

$$= 389.64658 \text{ pf (say:} \quad \text{CA} = 400 \text{ pf).} \quad (8.43)$$

$$\text{LB:} \quad \frac{L_2 R_L}{\omega_c} = \frac{0.9705 \times 240}{23.561945 \times 10^6} = \frac{2.2392 \times 10^{-6}}{0.23561945}$$

$$= 9.8854339 \text{ } \mu\text{h} \quad \text{(say:} \quad \text{LB} = 10 \text{ } \mu\text{h).} \quad (8.44)$$

$$\text{CC:} \quad \frac{C_1}{\omega_c R_L} = \frac{0.3374}{23.561945 \times 10^6 \times 240} = \frac{0.3374 \times 10^{-12}}{0.0056548668} = 60 \text{ pf} \quad (8.45)$$

Using the values (8.43), (8.44) and (8.45) in Fig. 8.16b., we obtain the unnormalized filter shown in Fig. 8.16c.

To obtain the differential configuration of the single-ended filter shown in Fig. 8.16c. it becomes necessary to draw the same network twice, in such a form that the lower network is the "mirror" image of the top network, as shown in Fig. 8.16d. Once this is done, the appropriate components can be combined, thus obtaining the final low-pass filter (differential form) shown in Fig. 8.16e.

8.6. DIFFERENTIATOR

In magnetic recording, the data is contained in the transitions, and this is the way it is recorded. However, when the data is read, the head (being an inductive element) takes the first derivative of the transition (Chapter 2). Therefore, now the data in the read waveform is contained in the peaks (not in the zero crossings, or transitions anymore). Consequently, it becomes necessary to obtain, again, the original information contained formerly in the transitions.

This can be done two different ways:

(a) *By integration* This would be equivalent to undoing the read head differentiation. Unfortunately, modern integrators offer an extremely high DC (plus low frequency) gains, they exacerbate the 1/f noise, they eliminate some useful high frequencies, etc. As a result, integration, although obvious, presents such formidable practical drawbacks that it has not been used in any disk configurations outside the development laboratory.

(b) *By differentiation* This means that the read head process is repeated. Although not quite straightforward, this has been the method of choice in practical disk drives. This method will be elaborated further.

Figure 8.17a. shows an active, differential differentiator implemented by means of an amplifier. The frequency response (in the Bode plot to the right) shows that the gain is directly proportional to the frequency. Obviously, this increases any remaining high-frequency noise that is intolerable. Therefore, let us consider the simplest low-pass filter possible, namely, an integrator, as shown in Fig. 8.17b. Since we saw already that the integrator should not be used alone, the best compromise would consist of combining Figs. 8.17a. and 8.17b., thus giving Fig. 8.17c. as the ideal frequency response sought.

A cursory view of Fig. 8.17c. shows that the desired frequency response looks very similar to the band-pass characteristics of a simple LC tank circuit. Therefore, it seems rather extravagant to use a circuit as expensive and complicated as the one shown in Fig. 8.17c. when for our case a much simpler circuit would be sufficient.

Figure 8.18a shows such a circuit. Figure 8.18b. shows the pole-zero characteristics desired and Fig. 8.18c. shows the required impedance. The conjugate poles are give by

$$-\frac{R}{2L} \pm j\frac{R}{2L}\sqrt{\frac{4}{R^2C}-1}, \qquad (8.46)$$

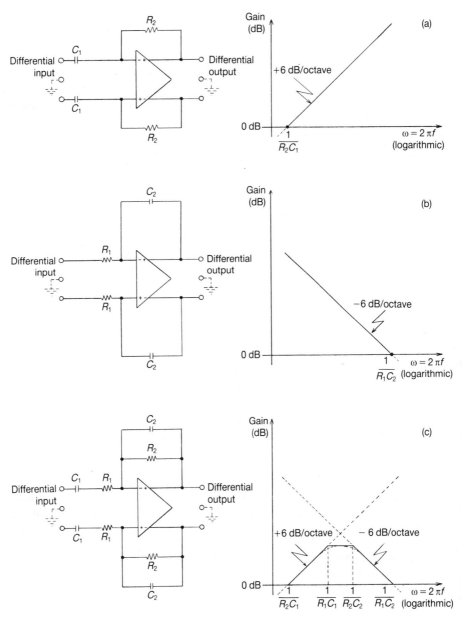

Fig. 8.17. Evolution of the differential active differentiator to the combined version. (a) Differentiator circuit and Bode Plot. (b) Integrator circuit and Bode Plot. (c) Combination circuit and Bode Plot.

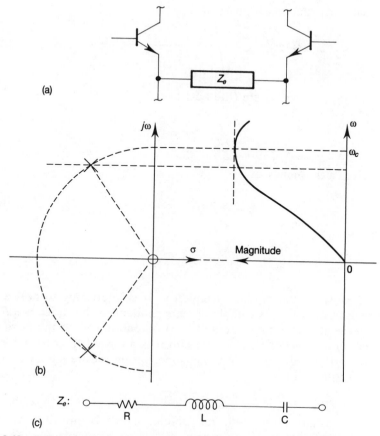

Fig. 8.18. Practical differentiator. (a) Functional impedance located between emitters. (b) Poles and zeros plus frequency response. (c) Passive components.

and the transfer function is given by

$$H(s) = \frac{A}{L} \left[\frac{s}{s^2 + \frac{R}{L}s + \frac{1}{LC}} \right],$$ (8.47)

where A = amplifier gain. Let us make $s = j\omega$ in (8.47), then

$$H(j\omega) = \frac{A}{L} \left[\frac{j\omega}{-\omega^2 + \frac{R}{L}j\omega + \frac{1}{LC}} \right] = \frac{A}{L} \left[\frac{j\omega}{\left(\frac{1}{LC} - \omega^2\right) + j\frac{R}{L}\omega} \right].$$

The magnitude function is given by

$$|H(j\omega)| = \frac{A}{L} \sqrt{\frac{\omega^2}{\left(\frac{1}{LC} - \omega^2\right)^2 + \left(\frac{R}{L}\omega\right)^2}}. \tag{8.48}$$

In Eq. (8.48), the maximum (the peak) occurs for

$$\frac{1}{LC} - \omega^2 = 0$$

which gives a frequency of

$$\omega_r = 2\pi f_b = \frac{1}{\sqrt{LC}}. \tag{8.49}$$

That is

$$f_b = \frac{1}{2\pi \sqrt{LC}},$$

where f_b is called the "break" frequency. It is interesting to notice that this break frequency f_b is completely independent of the resistance R.

The resistance R is implemented by a potentiometer at the beginning of a disk drive development effort, and it is adjusted to correct for any phase distortions remaining in the channel within the frequency range of interest, namely, for

$$F_L \leq f \leq F_H. \tag{8.50}$$

After the design is finalized, the potentiometer is substituted by an inexpensive resistor.

Example. Let us consider the same drive used for the low-pass filter, namely

Data Rate: $\Delta_B = 7.5$ megabits/sec, RLL(2, 7) code.

The rule of thumb is: For the differentiator break frequency, use a factor of 3.375 times F_H. For this case, we have

$$f_b = 3.375 \times F_H. \tag{8.51}$$

Substituting (8.42) in (8.51)

$$f_b = 3.375 \times 2.5 \text{ MHz} = 8.4375 \text{ MHz} \quad \text{and}$$

$$\omega_r = 2\pi f_b = 2\pi \times 8.4375 \times 10^6 = 53.014376 \times 10^6 \text{ radians/sec}. \tag{8.52}$$

Let us choose

$$L = 10 \; \mu h, \tag{8.53}$$

Substituting (8.43) and (8.44) in (8.49) and rearranging terms

$$C = \frac{1}{\omega_r^2 L} = \frac{1}{(53.014376 \times 10^6)^2 \times 10 \times 10^{-6}}$$

$$= \frac{10^{-12}}{0.028105241} = 35.58 \, \text{pf (say: } C = 36 \, \text{pf)}.$$

Note: Neither of the two rules of thumb factors given (one for the low-pass filter, the other for the differentiator) are rigorous. They are based on practical experience and could be changed in any specific application. However, they offer good starting points and might be changed later on, although they could be acceptable the first time.

8.7. DATA SEPARATOR

The "data separation" function sometimes is included in the controller and other times in the disk drive. In either case, that choice is usually one of convenience and/or packaging. However, in reality, it is an integral part of the read channel, therefore, it will be mentioned here.

There are two distinct cases to be considered:

(a) *Self-clocking codes (see Chapter 6)* For this particular case, the codes are required to have a (mandatory) clock pulse for every bit interval. Consequently, a function is needed to "separate" the data from the clock, thus giving two distinct waveforms from a single one, as shown schematically in Fig. 8.19. Usually, the data extraction portion also

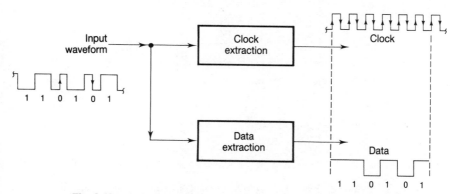

Fig. 8.19. Block diagram for a self-clocking code data separator.

performs a data conversion to the NRZ code which is used for logic implementation (Boolean algebra).

(b) *Other codes* (*not self-clocking*) For these codes the "data separation" function is no longer necessary, simply because there is absolutely nothing to separate and the clock is "implicit" in the recording waveform rather than physically present. For these codes, the clock must be "generated" rather than "separated."

Unfortunately, the term "data separator" has persisted because the self-clocking codes were used first. However, with the newer codes used today (such as the RLL (2, 7), see Chapter 7), it has become a misnomer and the creator of unnecessary confusion. That is, for these codes the so-called "data separator" does not "separate" anything anymore; instead, it performs the following functions:

(1) Generation of the clock signal,
(2) Amplification of the data stream,
(3) Synchronization between the two waveforms mentioned above, and
(4) Conversion of the data into NRZ form (however, usually it is better to perform this operation by means of a separate ENDEC IC chip).

A typical block diagram for this case is shown in Fig. 8.20. Obviously, a more appropriate name could be "clock generation and synchronization."

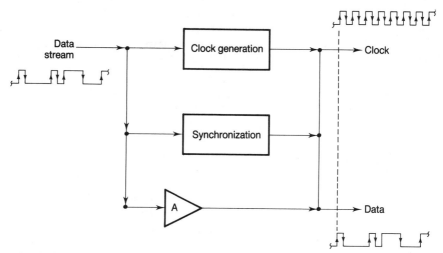

Fig. 8.20. Block diagram of a so called "data separator" for a code that is not self-clocking.

Many techniques have been used for the clock generation with various degrees of success. Among them:

(a) *One-shots* They are useful for FM (self-clocking code) where the clock frequency is always known. However, one-shots cannot track variations in the data rate. Consequently, they have been used only in very cheap low-end floppy disks.

(b) *Surface acoustic wave (SAW) filters* In this case, the Fourier Transform fundamental frequency is extracted by a band-pass filter. Variations in data rate have been followed by majority voting among several filters. SAW filters do not offer much flexibility because we are at the mercy of the specialized filter designer and also the state-of-the-art at the time of application.

The background of SAW filters is as follows. In the 19th century, Lord Rayleigh established the properties of surface waves in the frequency range of terrestrial seismic waves (Rayleigh waves). The invention of the interdigital coupler was the key to extending surface acoustic waves toward radio frequencies. Consequently, these filters are no longer limited to the rather narrow range of acoustic frequencies, namely, 20 Hz to 20 KHz.

These waves are a form of disturbance involving deformations of the material and the restoring forces that result. The material is described as "elastic" and the waves are often called "elastic waves." The technique of varying the electrode lengths is known as "apodisation."

The SAW filters have been used mostly as band-pass filters in delay lines, resonators, TV, radar, communication, etc. However, because of the lack of a DC path they cannot be used as low-pass filters.

Even today, their range of operation is quite narrow: In the time domain, they can be used within a delay of 0.1 microseconds to 50 microseconds. In the frequency domain, they can be used from 100 KHz (although at low frequencies they become quite bulky and should never be used below a few MHz) to 1.5 GHz, being determined by the present limitations of current lithography as developed for microelectronics.

New piezoelectric materials are being developed constantly as surface wave substrates, typically 1 mm thick. Although they are commercially attractive, their insertion loss ranges between 15 dB and 30 dB. For this and other reasons they have not been used extensively in disk drives. Figure 8.21. shows several typical SAW filters (Morgan, 1985).

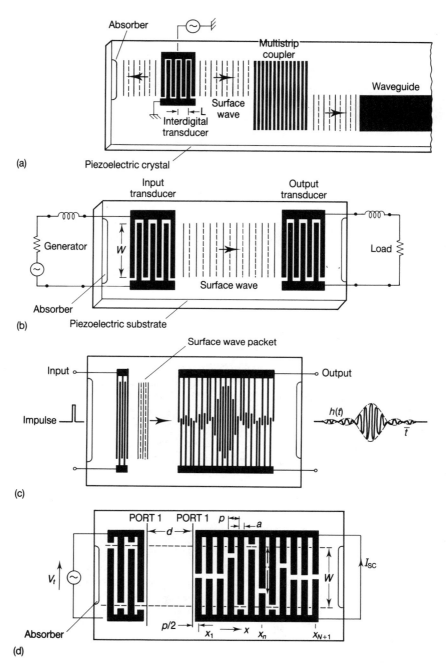

Fig. 8.21. Typical SAW filters. (a) Metal film components using the piezoelectric effect. (b) Interdigital delay line. (c) Bandpass filter using apodised transducer. (d) Bandpass filter using apodised transducer with regular electrodes. Reprinted with permission from *Surface-Wave Devices for Signal Processing,* David R. Morgan, Elsevier Science Publishers B.-V., 1985.

180

(c) *Digital phase locked loops* In reality, they can be considered high-speed synchronizers and can track variations in data rate. However, they require an internal clock at a high multiple of the data rate. As a result, they have been used mostly in floppy disks with data rates of less than 1 MHz.

(d) *Analog phase locked loops (PLL)* They consist of a servo loop in which a voltage controlled oscillator (VCO) is set to some frequency. This center frequency is modified (up or down) by a control voltage which represents the difference between the input (the data stream) and the VCO frequency. Obviously, within the "catch" range it can track variations in the data rate. This has become the method of choice for most hard disk drives. Also, it can be very flexible if we use an IC chip with the setting components mounted in an external hybrid package.

Figure 8.22. shows a simplified block diagram of a typical PLL. Its functioning is as follows: The phase detector compares the phase of an input signal against the phase of the voltage controlled oscillator (VCO). The output of the phase detector is directly proportional to the phase difference between its two inputs.

This difference voltage is sent to a low-pass filter (loop filter) whose transfer function is $F(S)$. The output of this filter is amplified and applied to the VCO, thus becoming the control voltage of the VCO. This control voltage changes the center frequency of the VCO in a direction (up or down) which tends to reduce the phase difference.

When the loop is "locked," the control voltage is such that the frequency of the VCO is exactly equal to the average frequency of the input signal. As previously stated, this has become the method of choice for most hard disk drives. Figure 8.22. shows the center frequency of the VCO being determined by a crystal. However, for relatively low frequency applications a simple capacitor is sometimes sufficient.

There exist in the marketplace a great number of integrated circuit PLLs which have been used together with discrete components to

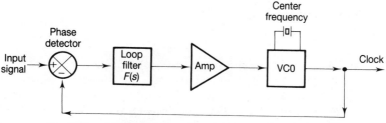

Fig. 8.22. Simplified block diagram of a phase-locked-loop.

implement the "data separation" function. At the same time, several companies are offering IC chips designed specifically for disk drives. In this case, those specialized chips also contain other functions that are needed in disk drives, among them:

(1) De-glitching,
(2) Read-enable-disable,
(3) Preamble detector,
(4) PLL lock detector, and
(5) Missing clock detector.

8.8. INTERFACE

In a data processing environment, there are two interfaces to consider:

(a) Interface between the central processing unit (CPU) and the controller, and
(b) Interface between the controller and the disk drive.

In this book we are not concerned with (a) above, only with (b). This is a subject that has become very extensive and complicated, and deserves a book of its own. There has been a tremendous proliferation of interfaces, among them:

ST 506/412,

ST 412 HP,

SASI/SCSI, and

ESDI.

Whichever the name of the interface, the signals can be classified into two distinct kinds:

(a) Control, and
(b) Data.

Since all the disk drives require (and produce) a finite number of signals, all interfaces, although seemingly different, are basically alike. Some provide (and require) more features than others. One of the simplest ones is the Enhanced Small Disk Interface (ESDI) illustrated in Figs. 8.23. and 8.24. and is typical of the interfaces being used today.

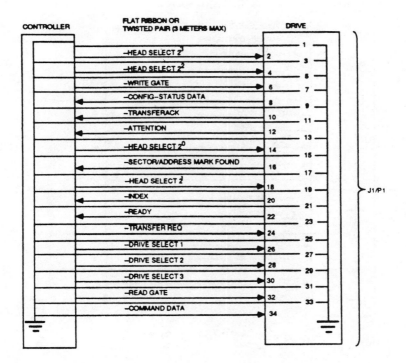

Fig. 8.23. ESDI control cable. (a) Signals. (b) Pin assignments. © 1989, Maxtor Corporation.

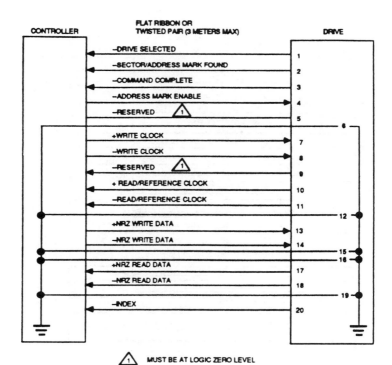

Fig. 8.24. ESDI data cable. (a) Signals. (b) Pin assignments. © 1989, Maxtor Corporation.

8.9. SUMMARY

This chapter has described the circuitry (R/W channel) needed to store and retrieve data in a DASD implemented by means of magnetic recording. However, the example above is still quite useful regardless of the recording phenomenon being used. That is, most of that circuitry will be necessary, consequently, the explanation above also constitutes a good foundation for any other method of recording used in the future. The most critical circuit components have been treated in more detail simply to emphasize their importance in a DASD environment.

9

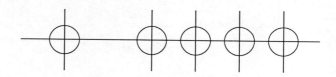

Access
Servomechanism

Before either the read or write process can commence, it is necessary to place the corresponding Read/Write (R/W) head on the appropriate track. The command to move the head to the proper track must be initiated by the computer. This command is interpreted by the controller, and the whole assembly of R/W heads, sliders, flexures, and block are moved in the radial direction of the disk until the desired track has been found. This mechanical movement is performed by means of a servomechanism. The design and implementation of a good servomechanism constitutes a specialty, so one of the books in this series is devoted completely to it. An introductory book on DASDs would be incomplete, however, without mentioning the needed servomechanism. This chapter will offer an introduction to a few of the most salient servo features that must be taken into consideration when using DASDs.

9.0. SERVOMECHANISM

The Read/Write (R/W) heads are positioned on the required track by means of a servomechanism. This implies some form of feedback. Many

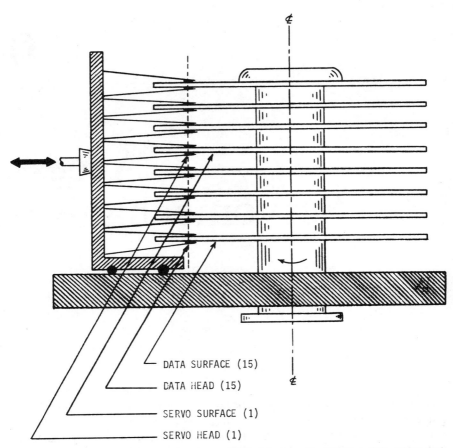

Fig. 9.1. Location of the servo surface and head in a multi–disk D.A.S.D. Illustrated for the case of 8 disks utilized as follows: 15 data surfaces plus one servo surface.

schemes use a dedicated servo surface, as shown in Fig. 9.1. The servo surface contains a great deal of information, i.e. location of the concentric tracks, index mark, etc. This information is written on the surface by the manufacturer with an extremely expensive and sophisticated device called the "servo-writer." Once this is done and the disk drive is assembled, tested, and delivered, the user cannot write on this surface; therefore, the servo head is a "read-only" head. Information obtained from this head is fed back and compared with the requested (by the controller) track number. For details on this operation, the reader is urged to consult the available literature (Oswald, 1974, 1978), although this subject is so extensive and specialized that it deserves a book of its own.

9.1. SPINDLE ROTATION

The servomechanism group in a disk drive design team is also usually assigned the task of controlling the spindle rotational speed. In many instances the spindle motor shaft contains three Hall switches, as shown in Fig. 9.2. The output from those switches offer a great deal of information, such as RPM, instantaneous shaft positioning, etc.

This information is fed to the spindle motor which controls that shaft, thus closing the loop, as shown in Fig. 9.3. (Kurzweil, 1987). Usually, this servo loop is very tight since the spindle rotational speed is kept within a very small tolerance such as $\pm 1\%$.

9.2. SERVO ACCURACY

A very good measure for a disk drive servomechanism is given by the number of tracks per inch T that the servo produces accurately and reliably. Figure 9.4. shows a graph of T versus years of first customer shipment (FCS) for several disk drives, past, present, and predicted future. Obviously, this number has been increasing at a tremendous pace.

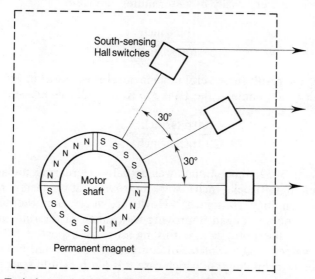

Fig. 9.2. Typical mounting of the three (3) Hall switches (sensors) in the spindle shaft of a rigid disk drive.

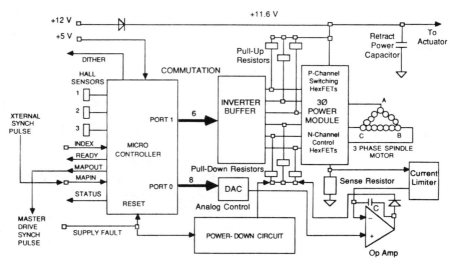

Fig. 9.3. Spindle rotational control. © 1989, Maxtor Corporation.

However, the magnetic recording density B in bits per inch (along a track) has been increasing at a tremendous pace also, and therefore, it becomes very interesting to determine which parameter has grown faster. This can be done very easily by determining the ratio

$$\frac{\text{bits/inch}}{\text{tracks/inch}}$$

and plotting the result for several machines. This is shown in Fig. 9.5. For the first disk drive, namely, the IBM 350 RAMAC, this measure was

$$\frac{100 \text{ bits/inch}}{20 \text{ tracks/inch}} = 5.$$

Until the year 1959, this amount was actually decreasing, meaning that servo technology (tracks/inch) was improving at a faster rate than recording technology (bits/inch). However, very soon thereafter, the recording technology began improving at a much faster rate than servo technology, to the point where this measure is expected to attain an asymptotic value of 22.5 (unless, of course, an unexpected breakthrough occurs in servo technology, in which case this figure would start becoming smaller again).

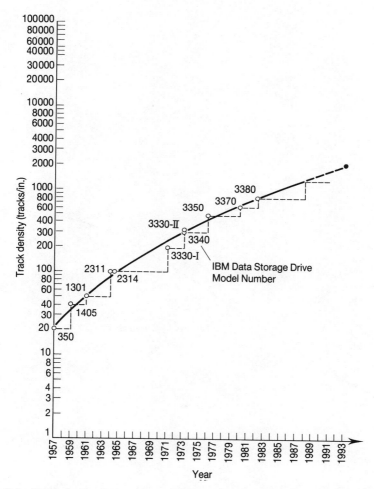

Fig. 9.4. Track density T in tracks per inch versus year of FCS. IBM machines only.

9.3. ACCESS TIME

Another very important figure of merit (for the servomechanism) is offered by the amount of time it takes the R/W heads to move from one track to another. Of course, it takes less time to move from one track to another that is adjacent than to a track that is relatively far away. Therefore, when talking "access time," we can refer only to averages, not absolutes.

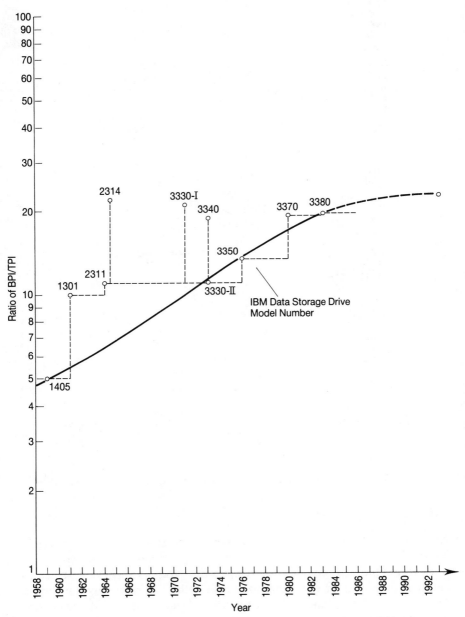

Fig. 9.5. Calculated and predicted ratio B/T versus year of FCS. IBM machines only.

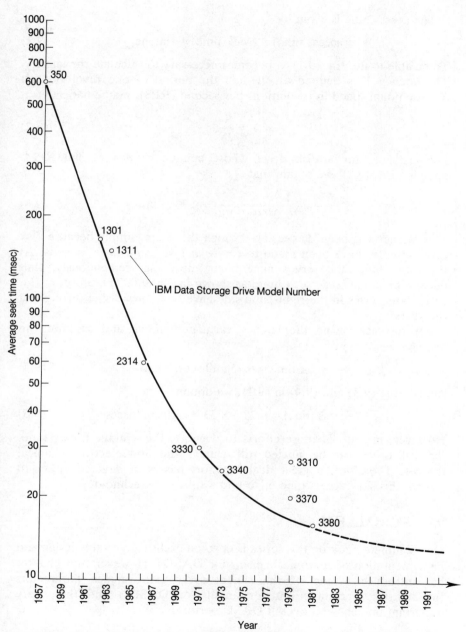

Fig. 9.6. Actual and predicted average seek time versus year of FCS. IBM machines only.

The access time is given by

$$\text{(access time)} = \text{(seek time)} + \text{(latency)} \tag{9.1}$$

To be able to use Eq. (9.1), it becomes necessary to calculate the latency. The latency Y is defined as one-half the period of one revolution. If S = rotational speed in revolutions per second (RPS), mathematically:

$$Y = \frac{1}{2S}. \tag{9.2}$$

For instance, for a disk drive whose spindle rotates at 3600 RPM, applying Eq. (9.2), we obtain that

$$Y = \frac{1}{2 \times (3600/60)} = \frac{1}{120} = 8.33 \text{ millisecs.} \tag{9.3}$$

This latency has been decreasing through the years, simply because disk drive spindles have been made to rotate at faster and faster speeds. At the same time, the servos have been improving tremendously, thus decreasing the average seek time, as shown in Fig. 9.6. Therefore, in Eq. (9.1), both terms in the right-hand side have been decreasing throughout the years.

For instance, using the latency calculated above and assuming an average seek time of

$$\text{seek time} = 6.75 \text{ millisecs,} \quad \text{and} \tag{9.4}$$

substituting (9.3) and (9.4) in (9.1), we obtain

$$\text{(access time)} = 6.75 + 8.33 = 15.08 \text{ millisecs} \tag{9.5}$$

However, in spite of those efforts to decrease the average access time, the disk drives can be labeled still, clumsy and slow electromechanical devices. Therefore, a great deal of effort has been devoted either to "bury" this slow access time or to find a substitute technology.

9.4. SUMMARY

One complete book on this series is devoted exclusively to the design and implementation of servomechanisms for DASDs. However, this chapter has introduced some of the salient servo performance parameters (needed for a complete understanding of DASDs) and their very important influence on overall DASD performance.

10

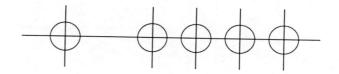

Optical versus Magnetic Recording

Due to the basic limitations of the magnetic recording process, many other phenomena of physics have been investigated and proposed (even implemented) as candidates for incorporation in DASDs. This chapter offers a mathematical comparison between optical and magnetic recording based exclusively on recording density (bits per square inch). Of course, there are many other parameters that must be taken into consideration before making a serious commitment. However, recording density is an extremely important performance parameter that seems to have been neglected in the literature on this subject. Of course, any new technology must be subjected to thorough scrutiny and to comparison with well–established (and known) technologies such as magnetic recording.

10.0. OPTICAL RECORDING

In an attempt to capture a share of the very large disk drive market, several technologies are being pursued that might be considered com-

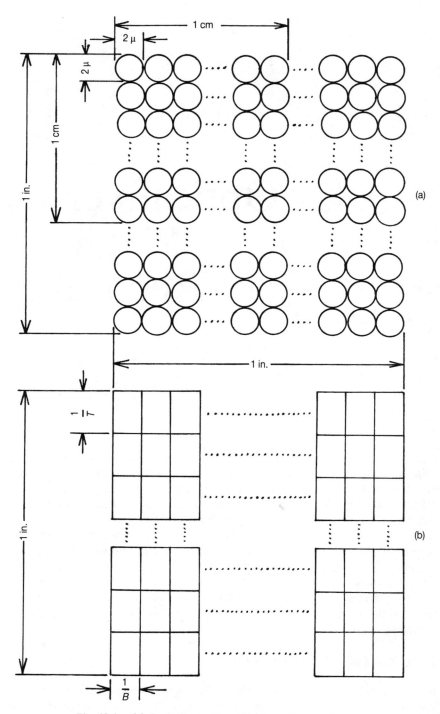

Fig. 10.1. (a) Optical recording. (b) Magnetic recording.

196

petitors of magnetic recording. The perennial candidate seems to be optical recording.

Let us refer to Fig. 10.1. One of the differences appears to be the shape of the recorded bit. Figure 10.1a. shows a circular (or elliptical) optically recorded bit with a 2 micron diameter (1 micron = 1-millionth of a meter). This is a dimension being pursued today, so for the sake of comparison, let us use it. The areal density A_o thus provided becomes:

$$A_o = (5{,}000) \times (5{,}000) = 25 \times 10^6 \text{ bits/square cm.} \tag{10.1}$$

It is known that:

$$1 \text{ inch} = 2.54 \text{ cm.} \tag{10.2}$$

Substituting (10.2) in (10.1)

$$A_o = 25 \times 2.54^2 \times 10^6 \text{ bits/square inch.} \tag{10.3}$$

Let us postulate a magnetic recording system, as shown in Fig. 10.1b.with B bits/linear inch (unknown) and T tracks/linear inch (also unknown). The areal density A_m is

$$A_m = B \times T. \tag{10.4}$$

It is interesting to find out what sort of magnetic recording system would give exactly the same areal density as the optical system considered above. Mathematically, let us make

$$A_m = A_o. \tag{10.5}$$

Substituting (10.4) and (10.5) in (10.3)

$$BT = 25 \times 2.54^2 \times 10^6. \tag{10.6}$$

Equation (10.6) consists of a single equation with two unknowns, namely, B and T. Therefore, it does not have a unique solution. To solve this equation uniquely, it becomes necessary to find a relationship between B and T. For this purpose, let us consider Fig. 10.2. which consists of a plot of the ratio $B \div T$ throughout the years. It is apparent that during our period of development, it is possible to establish that

$$B \div T = 22.5.$$

That is

$$B = 22.5 \, T. \tag{10.7}$$

Substituting (10.7) in (10.6)

$$22.5 \, T^2 = 25 \times 2.54^2 \times 10^6, \tag{10.8}$$

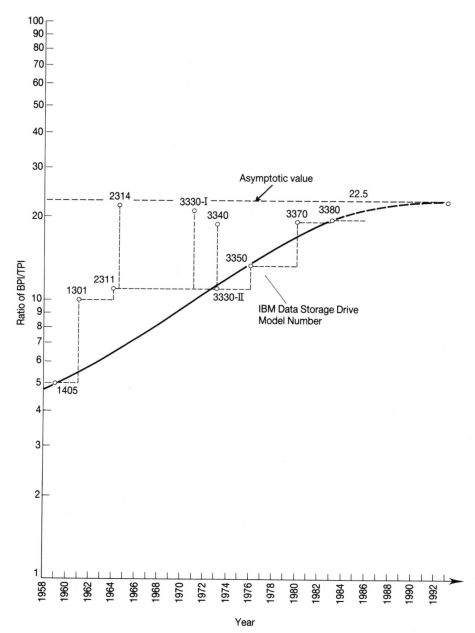

Fig. 10.2. Ratio of bits per inch to tracks per inch versus year of FCS for rigid disk drives.

whose solution is

$$T = 2,677 \text{ tracks/inch.} \tag{10.9}$$

Substituting (10.9) in (10.7)

$$B = 22.5 \times 2,677 = 60,233 \text{ bits/inch.} \tag{10.10}$$

substituting (10.9) and (10.10) in (10.4)

$$A_m = 60,233 \times 2,677 = 161.244 \times 10^6 \text{ bits/square inch.} \tag{10.11}$$

Equation (10.9) has been plotted in Fig. 10.3. Although the point of

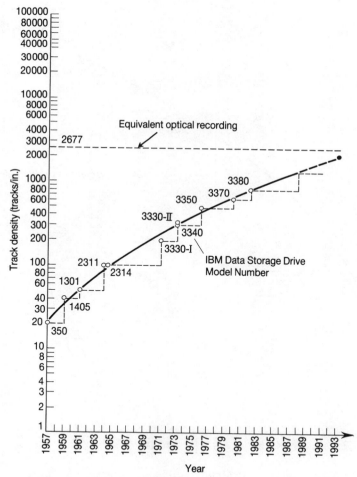

Fig. 10.3. Tracks per inch of optical versus magnetic recording.

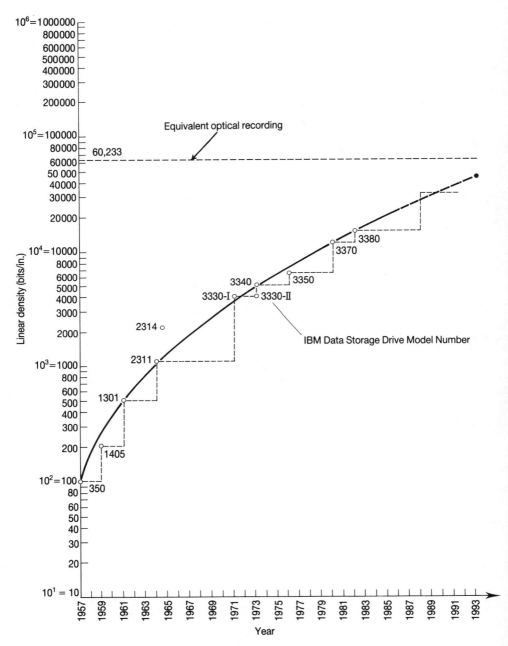

Fig. 10.4.　Bits per inch of optical versus magnetic recording.

coincidence falls outside our graph, we can see that a magnetic recording disk drive with a linear track density of 2677 tracks/inch will have a first customer shipment (FCS) around the year 1996.

Equation (10.10) has been plotted in Fig. 10.4. Although the point of coincidence falls outside our graph, we can see that a magnetic recording

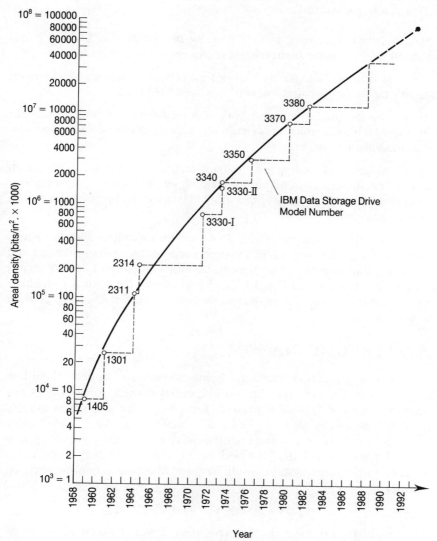

Fig. 10.5. Areal density of optical versus magnetic recording.

disk drive with a linear bit density of 60,233 bits/inch will have a FCS around the year 1996 also.

The number given in Eq. (10.11) falls outside our graph paper, as illustrated in Fig. 10.5. However, a simple extrapolation shows that a magnetic recording disk drive with an areal density of $161,244 \times 10^3$ bits/square inch will have a FCS around the year 1996.

We can make the following observations:

(A) Optical recording has a "window of opportunity" which is not infinitely wide. On the contrary, there are only a few years left.

(B) It is impossible to say that optical recording will never compete. This is due to the fact that "never" is a very long time.

(C) Our calculations show that optical recording has only a few more years in which to establish its credibility, unless, of course, an unforeseeable breakthrough occurs.

(D) Magnetic recording is a "moving target." Therefore, to mount an effective challenge, any competing technology must effect an immediate "quantum jump." Obviously, this does not seem to be happening with optical recording.

(E) As the optical recording spot becomes smaller than 2 micron diameter, the collimating optical system becomes more sophisticated and expensive, thus allowing more time to develop (more areal density means more years, as shown in Fig. 10.5). Unfortunately for optical recording, the price per megabit for magnetic recording keeps decreasing as time progresses.

10.1. SERVOMECHANISM

One of the greatest advantages being touted for optical recording happens to be the fact that the bits per inch are equal to the tracks per inch (for a ratio of 1), unlike magnetic recording (with an equivalent ratio of 22.5). Although, at first glance this fact seems like a tremendous advantage (because of the great areal density theoretically possible), in a practical sense this is a double-sided sword: From a random access point of view, our optical system would require a servomechanism with an accuracy of:

$$T = 2.54 \times 5000 = 12,700 \text{ tracks/inch.}$$

The state-of-the-art for rigid magnetic disk drives is about an order of magnitude smaller (1,500 tracks/inch).

A servomechanism capable of an accuracy of 12,700 tracks/inch, although not impossible, would require a very sophisticated and expensive double or triple servosystem which would prove uneconomical. Consequently, any current (and inexpensive) optical recording device would format the data as a single spiral instead of several concentric circles (not servo), thus eliminating the unique random access feature of present DASDs. Because the data is sequential, this product should be more of a competitor with tape drives.

10.2. RELIABILITY

Optical recording is being used today in compact disks and video disks (read-only applications). However, in the audio and video fields (entertainment), an error rate of 1 bit every 10^5 is quite acceptable. This is due to the fact that the human ear and retina accept this rate without problems because of the integrating (and compensating) capabilities of the brain. Unfortunately, present data processing (DP) devices (computers) are not that forgiving yet. As a consequence, present disk drives offer an error rate of 1 bit every 10^{12}; that is, a difference of seven orders of magnitude. This means that the two applications do not demand the same requirements. In fact, they are quite different. The DP field requires very stringent performance and much more than mere entertainment.

10.3. TRANSDUCERS

The transducers being used in compact and video disks happen to be rather inexpensive solid-state lasers. Magnetic recording employs R/W heads that are getting better, smaller, and less expensive all the time. In this regard (along with fiber optics, PIN diodes, flexures, sliders, etc.), the two technologies (optics and magnetics) are similar in the final price per megabit.

10.4. DEVELOPMENT

There are more than 70 companies in the world deeply involved in the development of disk drives. Additionally, magnetic recording technology (applied to disk and tape drives) has been under active and continuous development for more than 40 years. As a result, magnetic recording technology has involved thousands of people world-wide and the technology has been improved continuously. Unfortunately, optical technology for DP applications has not developed a similar "head of steam" yet.

10.5. CONCLUSIONS

(A)   Optical recording sounds very new although it has been around for years. Because of this mistaken notion, it gets very favorable press. As with a siren song, many unsuspecting developers are attracted to it and many patents are being issued. However, nothing practical (or economical) has come out of it to date. The potential payoff for DASD applications seems rather remote. Meanwhile, its development is expensive and time-consuming, so much so, that it is possible for a company to go bankrupt while chasing this rainbow hoping to find a pot of gold at the other end.

(B)   Optical recording (non-entertainment) has so many formidable drawbacks that many attempts have been made to find some immediate, special (through limited) applications for it. Among them we have:

(1)   WORM (Write Once Read Many). The market for this type of optical disk is limited to the narrow niches that can tolerate nonreversibility, such as archival storage. Little displacement of magnetic disk drives is foreseen for the future. Some displacement of tape in archival applications is probable.

(2)   CD-ROM. This type of technology benefits from industry agreement on the standards developed jointly by Sony (Japan) and Philips (Holland) for the entertainment industry. As with WORM, this type of technology has very limited application, and as of mid-1986, there were very few CD-ROMs installed and of them, none seemed to be a major success.

(C)   Reversible optical storage is the ultimate goal of the developers. Many basic principles of physics have been tried and discarded, new ones approached, and so on *ad infinitum* The problem has been narrowed down to one of searching for an appropriate material. Looking objectively, it seem that only a handful of very large organizations for example, the U. S. government, IBM, GE, AT&T, Sony, Fujitsu, Siemens, etc, can afford the staggering costs involved.

(D)   Other technologies. Figure 10.6. shows a summary of some of the technologies being pursued in the large memory market. This picture changes constantly, as old technologies are abandoned and new ones emerge. However, at this point in time, and for the foreseeable future, the winner seems to be magnetic recording.

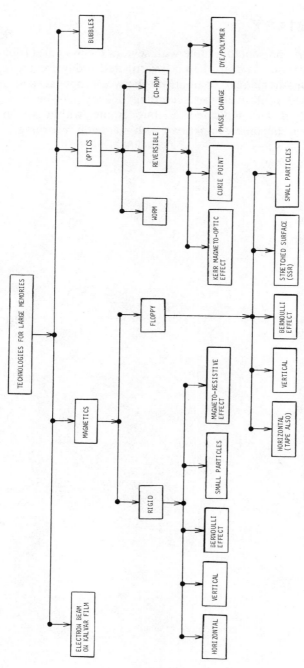

Fig. 10.6. Some technologies pursued for the large memory market.

205

10.6. SUMMARY

This chapter has presented a table with several technologies (not all) that have been proposed and are being investigated today for implementing DASDs. A mathematical comparison between optical and magnetic recording based exclusively on recording density has been presented. This comparison has indicated a time frame within which optical recording should become competitive with magnetic recording.

11

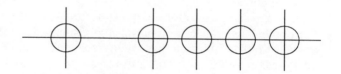

Array of
Disk
Drives

Presently implemented DASDs offer several important advantages:

(1) High capacity,
(2) Low cost per bit of storage, and
(3) Random access.

However, they offer also a few limitations:

(1) Low data rate (DASDs constitute rather slow electro-mechanical machinery),
(2) Low reliability (in spite of the progress that has been made), and
(3) Slow access to data (due to the rotational delay and slow radial movement of the Read/Write (R/W) heads).

Some of the disadvantages can be overcome at the system level by considering several DASDs, as a whole rather than as single entities. This can be done at the hardware and software levels as explained in this chapter, which falls more into the realm of computer science and data processing. To be sure, in the past, the total storage capacity has been increased by connecting several DASDs in a "daisy chain" configuration. However, when this is done, the data rate and reliability of the DASD being used at the moment become those of the total system. Another method (detailed in this chapter) consists of connecting the same devices in parallel (thus increasing the total data rate bandwidth) plus adding redundancy in the form of error correcting codes and extra DASDs (thus increasing the reliability of the overall system).

11.0. DISK ARRAYS

From 1977 to 1985, disk capacity has doubled every four years. However, during that same period of time, the disk market demand has doubled every two years. For high-capacity disk drives, this demand has been met primarily by single-spindle disk drives with 14-inch disks.

On the other hand, since 1985, the demand for disk drives has increasingly been met by disks of smaller diameters and correspondingly smaller capacity. Therefore, a question comes to mind immediately: Why not obtain high capacity by using many small-diameter disk drives? The answer to that question has been receiving a great deal of attention lately. Many other questions still remain unanswered, and a tremendous amount of system research is bring lavished on this subject. Consequently, in this book, it is only possible to mention some of the pertinent subjects with the hope of stimulating further inquiries.

11.1. SPINDLE SYNCHRONIZATION

When a large capacity disk drive is built by means of a single spindle, this spindle is made to rotate by means of a single motor that also rotates all the disks. There is a single index mark, and all the data written or read from the disk drive are referenced to this index mark.

However, when we are dealing with an assortment of disk drives, each separate disk drive has its own spindle, rotating motor, and index mark. The next question that arises becomes: Should they behave as if they had a single spindle? Or, expressing it another way, should they be synchronized?

Figure 11.1. is an illustration of the problem encountered. There are N disk drives, each rotating at the same S revolutions per second. When the spindles are not synchronized, we can expect that at any given point in time, their respective index marks are situated at random. However, when all the spindles are synchronized to a master drive, all the different

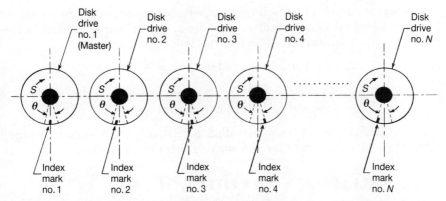

Fig. 11.1. *N* disk drives with synchronized spindles shown at a specific point in time.

index marks remain confined within a rather small angle θ of each other at all times. There are two factors to be considered:

(1) Time required to achieve synchronization, or acquisition time t_a, and
(2) Window width, or angle θ expressed in units of time, t_w.

Therefore, spindle synchronization has to do with the mechanical rotation of the spindles, plus mechanical vibrations, heat variations, etc. It is ironic that this basically mechanical problem presents a purely mechanical solution that is unthinkable. That is, a purely mechanical approach involving only gears, pulleys, belts, etc., would result in a very expensive and nightmarish contraption.

The only other practical approach would be electro-mechanical, that is, the use of electronic circuitry compensating for (not eliminating) the unwanted mechanical variations. Of course, we know that the best electro-mechanical devices employ servomechanisms. Therefore, a servo approach seems reasonable.

Most of today's disk drive rotational shafts have three Hall switches (sensors). Aside from all their other functions, the Hall sensors convert the shaft rotation to electrical signals. In this respect, they are transducers, that is, they effect the translation of mechanical motion into electrical signals. There is another transducer, namely, the servo read head which (among other things) translates the servo surface index mark into an electrical pulse. Thus, the transducers have effectively transferred a mechanical problem to the electrical realm.

The electrical pulses mentioned above can now be used to generate an error signal and also to synchronize other disk drives (to the first one, or master drive) by means of servo algorithms. One of the references in the Bibliography (Kurzweil, 1987) gives the following figures:

$$t_a = 1 \text{ second (from as far away as 8333 micro-}$$

$$\text{seconds), and}$$

$$t_w = 40 \text{ microseconds } (\pm 20 \text{ microseconds)}.$$

It is reasonable to expect that, within a relatively short period of time, these figures will be improved by making them even smaller.

11.2. RELIABILITY "BATHTUB" CURVE

Another matter to consider is the reliability implication of using an array of small disk drives as opposed to a single one. That is, we must know what reliability changes take place when we consider an array of small disk drives.

For this purpose, let us get started by considering basic reliability principles. For instance, three separate curves, as shown in Fig. 11.2. are used to characterize the failure rate of a product. Each curve corresponds to a failure mode type: (a) early failure, (b) random failure, and (c) wearout failure.

Reliability theory and practice differentiates between these different types of failures for the following reasons:

- They occur at different periods of time during a product's life cycle.

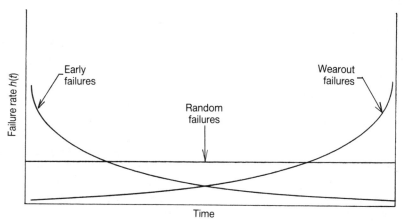

Fig. 11.2. The three curves that make up the reliability life curve.

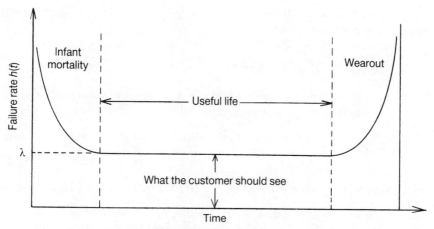

Fig. 11.3. The "bathtub" curve.

• Each curve has a different distribution and a different mathematical treatment.
• Different methods must be used for their prevention and/or reduction.

Figure 11.3. is the typical "bathtub" curve, which is nothing more than the result of the three curves of Fig. 11.2. summed.

During the useful life of a product, the hazard function $h(t)$ is a constant λ called the "failure rate." That is, for this interval, *only and exclusively* can we make the following assumption:

$$h(x) = \lambda = \text{constant} = \text{failure rate.} \tag{11.1}$$

Then, the reliability function $R(t)$ becomes

$$R(t) = e^{-\int_0^t \lambda \cdot dx} \tag{11.2}$$

which gives

$$R(t) = e^{-\lambda t} \tag{11.3}$$

where

$$\text{MTBF} = \int_0^\infty R(t)\, dt = \int_0^\infty e^{-\lambda t}\, dt;$$

$$\text{MTBF} = \frac{1}{\lambda}. \tag{11.4}$$

What the disk drive user would like to see is the flat portion of the "bathtub" curve which corresponds to this constant failure rate λ.

Consequently, during all the subsequent reliability calculations we will assume this condition to hold.

11.3. SERIES RELIABILITY

Let us consider a set of N events that are mutually exclusive, as shown in Fig. 11.4. There are two different probabilities to be considered:

$$P(A \text{ or } B \text{ or } C \text{ or} \ldots \text{ or } N) = P(A) + P(B) + P(C) + \ldots + P(N); \quad \text{and}$$
$$(11.5)$$

$$P(A \& B \& C \& \ldots \& N) = P(A) \times P(B) \times P(C) \times \ldots \times P(N). \quad (11.6)$$

This is the case when events $A, B, C, \ldots, N$ are statistically independent. Furthermore, let us consider the case when the individual events are so linked that if any of them fails, the whole system is considered failed. When this happens, from a statistical (and reliability) point of view, Eq. (11.6) applies, and the events (in our case, disk drives), can be considered in series, as shown in Fig. 11.5.

Note. The reliability diagrams (such as the one in Fig. 11.5) are concerned, exclusively, with the failure modes, not with the physical interconnection (usually the wiring diagram) of the units under consideration.

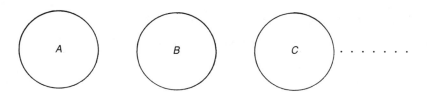

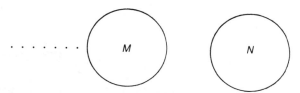

Fig. 11.4. A set of N events that are mutually exclusive (statistically independent).

Fig. 11.5. Reliability diagram for a series configuration (8 drives).

Since reliabilities are probabilities, we have

$$R_s[(\text{No. }1)\ \&\ (\text{No. }2)\ \&\ (\text{No. }3)\ \&\ \ldots\ \&\ (\text{No. }N)]$$

$$= (R_1) \times (R_2) \times (R_3) \times \ldots \times (R_N), \quad \text{where s = system.} \quad (11.7)$$

Substituting (11.3) in (11.7)

$$R_s = e^{-\lambda_s t} = (e^{-\lambda_1 t}) \times (e^{-\lambda_2 t}) \times (e^{-\lambda_3 t}) \times \ldots \times (e^{-\lambda_N t}), \quad (11.8)$$

taking natural logarithms, and simplifying

$$\lambda_s = \lambda_1 + \lambda_2 + \lambda_3 + \ldots + \lambda_N. \quad (11.9)$$

Applying Eq. (11.4) to (11.9),

$$(\text{MTBF})_s = \frac{1}{\lambda_s} = \frac{1}{\lambda_1 + \lambda_2 + \lambda_3 + \ldots + \lambda_N}. \quad (11.10)$$

For the particular case when

$$\lambda_1 = \lambda_2 = \lambda_3 = \ldots = \lambda_N = \lambda,$$

Eq. (11.10) simplifies to

$$(\text{MTBF})_s = \frac{1}{N\lambda}. \quad (11.11)$$

Example. Let us consider the specific case $N = 8$, that is, 8 equal models of disk drives operating so that each one of them produces a bit such that all of them taken together produce a parallel byte. In this case, if any one of the disk drives fails, the byte is no good and the whole system can be considered to have failed. For this case, Eq. (11.11) applies.

Let us consider a disk drive manufacturer that specifies an MTBF of 30,000 hours for each disk drive, that is, three years, plus 22 weeks, plus one day. Then

$$\lambda = \frac{1}{30,000} \text{ hours}^{-1}. \quad (11.12)$$

Substituting (11.12) in (11.11)

$$(\text{MTBF})_s = \frac{1}{8/30,000} = 30,000/8 = 3,750 \text{ hours},$$

which is equal to 22 weeks, plus two days, plus six hours.

In conclusion, without reliability improvements, large arrays of disk drives are too unreliable to be useful. Consequently, we have no other alternative but to find means to increase the extremely poor MTBFs.

11.4. RELIABILITY IMPROVEMENTS

The basic ways of improving reliability are

(a) *Lowering the parts count* implementing the function in another way, with fewer parts, or removing a circuit function.

(b) *Using higher grade parts* using the same type of part, but one that has undergone a better manufacturing and screening process so that it has a lower failure rate.

(c) *Reducing stress levels* reducing the voltage, current, power, and/or temperature. Also, reducing the use of the device by lowering its duty cycle.

(d) *Improving fault tolerance capability* there are three major design techniques:

Triple modular redundancy (TMR), or majority voting:

  (i)   Voting logic must be highly reliable.
  (ii) Advantages are that fault-masking occurs immediately, and both temporary and permanent faults are masked.
  (iii) Implementation problem is synchronization between modules (must use a common fault-tolerant clock).

Error-correcting codes:

  (i)   Used in computers to improve main memory reliability.
  (ii) Used in disk drives (usually by means of Fire codes) to improve R/W reliability (thus masking some disk area defects).
  (iii) Check bits are added to information bits in such a way that if errors occur, they can be detected (error detection) and some of the original data bits in error can be reconstructed as if no errors had occurred (error correction).

(iv) Several semiconductor manufacturers have introduced error detection and correction chips.

Redundancy.
There are three different types of redundancy to be considered:

(i) Time redundancy:
  - Commonly used in the detection and correction of errors caused by temporary faults.
  - Involves repetition or rollback of instructions or program segments immediately after a fault is detected.
  - Restarts processing from last check point.
  - If fault is temporary, rollback to a checkpoint should allow successful recovery.
  - If fault is permanent, detection mechanism will be activated again and an alternative recovery method (if possible) will be attempted.
  - Selection of correct number of checkpoints is important.
  - Too many checkpoints increase computation time.
  - If the checkpoints are too far apart, the recovery time will increase.
  - Already, most disk drive manufacturers include this technique in their products. Consequently, it is not necessary for us to delve into it anymore. We take it for granted that it exists (in one form or another) already in the disk drives used in our arrays.
(ii) Active Redundancy:
  - All redundant units are operating simultaneously rather than being switched on when needed.
  - Active parallel redundancy ($\lambda_1 = \lambda_2$).
  - Active parallel functional redundancy ($\lambda_1 \neq \lambda_2$).
  - Active partial redundancy (k out of n).
(iii) Standby Redundancy:
  - All standby redundant units are inoperative (dormant) until needed. Upon failure, the standby unit is powered on and/or switched to replace the primary unit.
  - Standby units alike ($\lambda_1 = \lambda_2$).
  - Functional redundant units ($\lambda_1 \neq \lambda_2$).

These methods of reliability improvement are summarized in Fig. 11.6. At this time, disk drive manufacturers employ most of these methods to improve the drive MTBF, which is presently quoted as 30,000 hours.

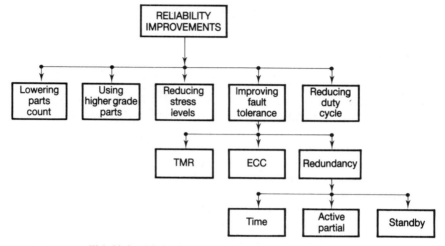

Fig. 11.6.   Methods used for reliability improvement.

Therefore, for the case modes consideration, namely, disk arrays, we are going to deal exclusively with the last cases, namely: (a) active partial redundancy, and (b) standby.

11.5. ACTIVE PARTIAL REDUNDANCY

For this case of partially redundant systems, successful operation is achieved if at least k units are operative out of n parallel units, otherwise the operation is a failure. In this case, we are dealing with a sequence of binomial trials, and we are interested in the probability of a number of successes. If we let the random variable k equal the number of desired successes in n trials, the probability of k successes in n trials is

$$P(k) = \binom{n}{k}p^k(1-p)^{(n-k)}. \tag{11.13}$$

The cumulative probability distribution function (CDF) represents the probabilities for k or more successes in n trials:

$$\text{CDF} = \sum_{i=k}^{n} \binom{n}{i}p^i(1-p)^{(n-i)}. \tag{11.14}$$

Therefore, in such cases, the reliability of the redundant group (the total system) is given by such a series of additive binomial terms. That is,

$$R(t) = \sum_{i=k}^{n} \binom{n}{i}p^i(1-p)^{(n-i)}. \tag{11.15}$$

For the exponential case, let us apply Eq. (11.3) to Eq. (11.15); we have that

$$R(t) = \sum_{i=k}^{n} \binom{n}{i} (e^{-\lambda t})^i (1 - e^{-\lambda t})^{(n-i)}, \tag{11.16}$$

where the range for $i = k$ happens to be

$$k, \ldots, (n-3), (n-2), (n-1), n. \tag{11.17}$$

The reliability diagram applicable to this case is shown in Fig. 11.7.

Case 1. No Reconstruction of Failed Drive. In this case, since drive failures are not being repaired, we should use MTTF, which is the average time to the first failure only (in cases to first failure, we should not use MTBF). That is

$$(\text{MTTF})_s = \int_0^\infty R_s(t) \, dt. \tag{11.18}$$

Substituting (11.16) in (11.18)

$$(\text{MTTF})_s = \int_0^\infty \left[\sum_{i=k}^{n} \binom{n}{i} (e^{-\lambda t})^i (1 - e^{-\lambda t})^{(n-i)} \right] \cdot dt$$

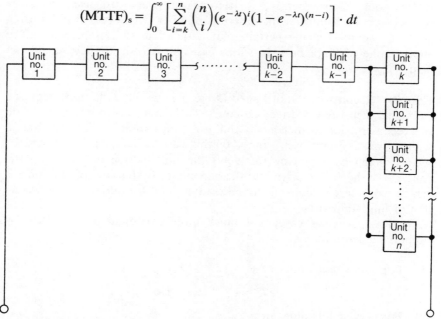

Fig. 11.7. Reliability diagram for active partial redundancy (k out of n), or at least k units operating out of n.

which simplifies to

$$(\text{MTTF})_s = \frac{1}{\lambda}\left[\sum_{i=k}^{n}\left(\frac{1}{i}\right)\right] \tag{11.19}$$

Example. Let us apply Eq. (11.19) to the case of nine drives producing one eight bit byte in parallel. For this case, we have

$$k = 8, \qquad n = 9.$$

Applying these values and Eq. (11.12) to Eq. (11.19)

$$(\text{MTTF})_s = \frac{1}{\dfrac{1}{30,000}}\left[\frac{1}{8} + \frac{1}{9}\right] = 30,000 \times \left[\frac{17}{72}\right];$$

$$(\text{MTTF})_s = 7,083 \text{ hours} \tag{11.20}$$

which is equal to 42 weeks, plus three days, plus three hours.

In conclusion, it is obvious that adding an extra drive (as opposed to the series case where we had only eight drives) has increased the MTTF from 3750 hours to 7083 hours. Although an improvement, it is still rather poor. Consequently, it becomes advisable to use an error-correcting code to further increase this number. However, even by using an ECC we must make certain that a failed drive is replaced before another one of them fails, in which case the existing ECC's would not help.

Case 2. Reconstruction of Failed Drive. Let us call T the time required to reconstruct a failed drive. Of course, it is much more desirable to do this reconstruction in such a way that it is transparent to the user, that is, without wasting any user's time. In this case, a failed drive should be reconstructed by replacing (actively) the failed drive and copying the contents of the failing drive into the new drive. In this context, the time T can be considered "copying time," rather than the physical reconstruction of the failing drive.

For this particular case, it is much more expedient to make several simplifying assumptions:

(1) For very small values of x,

$$e^{-x} \approx 1 \qquad (\text{for} \quad x \ll 1).$$

(2) For very small values of x

$$1 - e^{-x} \approx x \qquad (\text{for} \quad x \ll 1).$$

(3) Regardless of the size of T, the system failure rate is a constant, which means that the exponential distribution holds (at all times) for the whole array of disk drives.

For this case, it is much more convenient to use one particular reference in the Bibliography (O'Connor, 1985, Eq. (5.5), p. 131) applied to the exponential case. That is

$$R(t) = 1 - \sum_{i=0}^{k-1} \binom{n}{i} (e^{-\lambda t})^i (1 - e^{-\lambda t})^{(n-i)}. \qquad (11.21)$$

Applying (11.21) to

$$t = T,$$

we have

$$R(T) = 1 - \sum_{i=0}^{k-1} \binom{n}{i} (e^{-\lambda T})^i (1 - e^{-\lambda T})^{(n-i)}. \qquad (11.22)$$

Applying our first two assumptions, Eq. (11.22) simplifies to

$$R(T) = 1 - \sum_{i=0}^{k-1} \binom{n}{i} (\lambda T)^{(n-i)}. \qquad (11.23)$$

Note: In this case, our first two assumptions above are quite applicable. for instance, according to Eq. (11.12), we are given by the disk drive manufacturers that

$$\lambda = \frac{1}{30,000} \text{ hours}^{-1}.$$

Therefore, even if we consider a rather large value for T, such as

$$T = 10 \text{ hours},$$

we still have that

$$\lambda T = \frac{10}{30,000} = 0.00033333,$$

which is, indeed, much smaller than 1. Another way of corroborating those two assumptions above, is by verifying that

$$e^{-0.00033333} = 0.99966667 \approx 1 \quad \text{(first assumption); and}$$

$$1 - e^{-0.00033333} = 0.00033333 \quad \text{(second assumption)}.$$

But by means of our third assumption above, we have that

$$R(T) = e^{-\lambda_s T}$$

which can be approximated by

$$R(T) \cong 1 - \lambda_s T. \tag{11.24}$$

Substituting (11.24) in (11.23)

$$1 - \lambda_s T = 1 - \sum_{i=0}^{k-1} \binom{n}{i} (\lambda T)^{(n-i)},$$

which gives

$$\lambda_s = \frac{1}{T} \left[\sum_{i=0}^{k-1} \binom{n}{i} (\lambda T)^{(n-i)} \right]. \tag{11.25}$$

Applying (11.4) to (11.25)

$$(\text{MTBF})_s = \frac{T}{\sum_{i=0}^{k-1} \binom{n}{i} (\lambda T)^{(n-i)}} \tag{11.26}$$

Example.   Let us apply Eq (11.26) to a disk array with

$$n = 9, \quad k = 8.$$

Substituting these values in (11.26)

$$(\text{MTBF}) = \frac{T}{\binom{9}{0}(\lambda T)^9 + \binom{9}{1}(\lambda T)^8 + \binom{9}{2}(\lambda T)^7 + \binom{9}{3}(\lambda T)^6 + \binom{9}{4}(\lambda T)^5}$$

$$+ \frac{T}{\binom{9}{5}(\lambda T)^4 + \binom{9}{6}(\lambda T)^3 + \binom{9}{7}(\lambda T)^2}; \tag{11.27}$$

$$(\text{MTBF})_s = \frac{T}{(\lambda T)^9 + 9(\lambda T)^8 + 36(\lambda T)^7 + 84(\lambda T)^6 + 126(\lambda T)^5}$$

$$+ \frac{T}{126(\lambda T)^4 + 84(\lambda T)^3 + 36(\lambda T)^2}.$$

Let us simplify Eq. (11.27) by keeping only the largest term in the denominator. We obtain

$$(\text{MTBF})_s \cong \frac{T}{36\lambda^2 T^2} = \frac{1}{36\lambda^2 T}. \tag{11.28}$$

Substituting (11.12) in (11.28)

$$(\text{MTBF})_s \cong \frac{(30,000)^2}{36T} = \frac{900 \times 10^6}{36T}. \quad \text{hours}$$

Finally,

$$(MTBF)_s \cong \frac{25 \times 10^6}{T} \text{ hours.} \tag{11.29}$$

Therefore, the system reliability is inversely proportional to the reconstruction T and it becomes advantageous to keep T as small as possible. Also, it is advisable to include reconstruction time T as an integral (and important) part of any disk array. Equation (11.29) has been plotted in a log-log paper and shown in Fig. 11.8. Also, several pertinent values for T have been tabulated in Table 11.1.

It is interesting to find the amount of time T_e required to reconstruct a

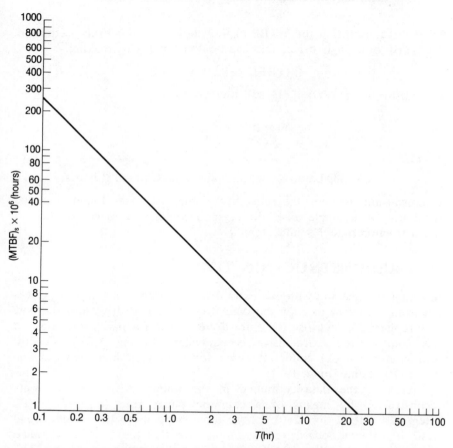

Fig. 11.8. Plot of system MTBF versus drive reconstruction time T for 8 out of 9 drives, active redundancy.

Table 11.1

A Few System MTBFs versus Reconstruction Time T for Eight out
of Nine Disk Drives, Active Redundancy.

T (Hours)	$(MTBF)_s$ (Hours and calendar nomenclature)
0.1	$250 \times 10^6 = 28{,}538$ years, 42 weeks, 2 days, 16 hours
1	$25 \times 10^6 = 2{,}853$ years, 42 weeks, 6 days
10	$2.5 \times 10^6 = 285$ years, 20 weeks, 1 day, 16 hours
100	$0.25 \times 10^6 = 28$ years, 28 weeks, 16 hours
1000	$0.025 \times 10^6 = 2$ years, 44 weeks, 3 days, 16 hours

failed drive, such that the MTBF of the whole system is exactly equal to
the MTBF of a single drive. This can be done simply by making

$$(MTBF)_s = 30{,}000 \text{ hours} \qquad (11.30)$$

and substituting (11.30) in (11.29), thus giving

$$30 \times 10^3 = \frac{25 \times 10^6}{T_e}.$$

That is

$$T_e = 833.333 \text{ hours} = 4 \text{ weeks, plus 6 days, plus } 17\tfrac{1}{3} \text{ hours.}$$

In conclusion, to keep the reliability of the disk array better than the
reliability of a single disk drive, it becomes necessary to have a
reconstruction time T smaller than T_e.

11.6. RECONSTRUCTION TIME

In actual practice and with one extra drive for redundancy purposes, it is
not even necessary to copy the data from the failed drive into the new
onw replacing it, because the extra drive is used for parity purposes. In
this context, the extra drive is equivalent to the VRC (Vertical
Redundancy Check) commonly used for nine-track high density tape
drives (Patel and Hong, 1974).

Therefore, the data contained in the failing drive can be easily
reconstructed by means of a common parity tree (e.g., the TTL
Integrated Circuit SN74280), whose input happens to be the data read
from the other (good) drives (assuming now that the data in the failed
drive consisted of all zeros thus not affecting the new parity count). The
parity thus obtained from the parity tree (either one or zero) is written on

the "hot spare." This is done byte by byte. Of course, the total procedure can be done just as easily by means of programming.

As stated before, the system MTBF is inversely proportional to the reconstruction time T. Consequently, this time T must be made as small as possible. Figure 11.9. Shows a flow diagram of an algorithm for accomplishing this. The method is as follows:

(a) Start reconstructing the data (contained in the lost, failed drive) by reading from the outermost (or innermost) track of all other non-failed drives.
(b) For the time being, assume that it takes only one revolution to reconstruct each track.

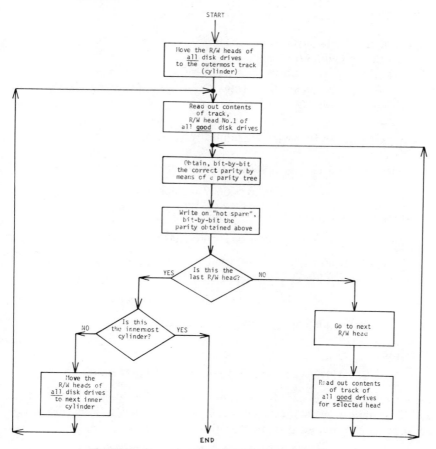

Fig. 11.9. Reconstruction procedure. Flow diagram.

(c) Using the parity obtained from the remaining good drives (EITHER by hardware or software) write the appropriate one or zero on the "hot spare."

(d) This way, a whole cylinder can be reconstructed (without wasting any time) simply by switching electronically from R/W head to R/W head (which is equivalent to a reconstruction from surface to surface).

(e) When finished with the whole cylinder, move all the R/W heads to the next (adjacent) cylinder and towards the center (or from the center) of the disk drives (all of them).

Example. The following disk drive parameters are offered by a manufacturer:

(1) Number of data surfaces: 15.
(2) Number of cylinders: 1,632.
(3) Period of one revolution: 16.666667 millisec.
(4) Track-to-track seek time (maximum and including R/W head settling): 2 millisec.

We can perform the following calculations:

(1) Time required for the revolution of all cylinders

$$= 15 \times 16.666667 = 250 \text{ millisec.}$$

(2) R/W time for all cylinders

$$= 250 \times 1{,}632 = 408{,}000 \text{ millisec.}$$

(3) Seek time from track to track (which in this case is the same as from cylinder to cylinder)

$$= 2 \times 1{,}632 = 3{,}264 \text{ millisec.}$$

(4) Total time $= 408{,}000 + 3{,}264 = 411{,}264$ millisecs.

Therefore,

$$T = 411.264 \text{ sec} = 6.8544 \text{ minutes} = 0.11424 \text{ hours.}$$

However, this figure can be considered rather optimistic. In a more realistic situation, we can assume that it takes two revolutions (instead of only) to reconstruct each track. For this case,

$$\text{R/W time for all cylinders} = 2 \times 408{,}000 = 816{,}000 \text{ millisec.}$$

Therefore,

$$T = 819.264 \sec = 13.6544 \text{ minutes} = 0.2275733 \text{ hours.} \quad (11.31)$$

For the case of standby redundancy (to be covered in the next section), we have that

$$(\text{MTBF})_s = \frac{50 \times 10^6}{0.2275733} = 219,709,430 \text{ hours,}$$

(which is equivalent to 25,080 years, plus 51 weeks, plus 2 days, plus 14 hours).

11.7. STANDBY REDUNDANCY

Standby redundancy is achieved when one unit does not operate continuously but is switched on only when the primary unit fails. A standby electrical generating system is an example. Figure 11.10. is a reliability diagram for a very simple case of this type of redundancy. The standby redundancy unit is sometimes called the "hot spare."

The sensing and switching system may be considered to have a hazard rate λ_s. Also, the switch and redundant unit may have a dormant hazard rate λ_d, particularly if they are not maintained or monitored. However, for purposes of simplification, we are going to assume

$$\lambda_s \cong \lambda_d \cong 0. \quad (11.32)$$

That is, for this simple standby model, we are going to assume that the sensing (monitor) and switching mechanisms are 100 % reliable and that the dormant failure rate is zero.

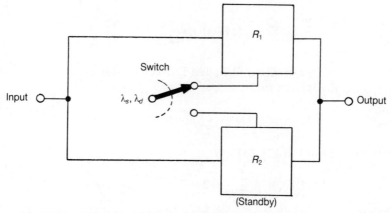

Fig. 11.10. Reliability diagram for a simple case of standby redundancy.

Also, we are going to assume that

$$R_1 = R_2. \tag{11.33}$$

For our prticular case, this last assumption is not unreasonable simply because when one disk drive fails (is removed) and the "hot spare" switched in, reconstructed, and made operational, the reliability of the new system should be exactly the same as the previous reliability (before the switching took place). That is, the system MTBF before and after the switching should be the same, although only one drive, and not all of them was switched in.

Formula derivation. There are only two ways in which successful operation can be achieved:

(1) Unit 1 survives during the full operation time t; or
(2) Unit 1 fails at time t_1 (prior to the end of the operating time t), and then Unit 2 takes over (after switching) and completes the job until the full operating time t.

For the exponential case the total reliability is given by

$$R(t) = e^{-\lambda t} + \lambda t e^{-\lambda t}. \tag{11.34}$$

Then

$$(\text{MTBF})_s = \int_0^\infty R(t) \cdot dt = \int_0^\infty e^{-\lambda t}\, dt + \int_0^\infty \lambda t e^{-\lambda t}\, dt$$

$$= \frac{1}{-\lambda}(0 - 1) + \lambda\left(\frac{1!}{\lambda^2}\right) = \frac{1}{\lambda} + \frac{1}{\lambda};$$

$$(\text{MTBF})_s = \frac{2}{\lambda}. \tag{11.35}$$

Case 1. No Reconstruction of Failed Drive. For this case, we calculated already (see Eq. 11.20) that

$$\lambda = \frac{1}{7,083}\ \text{hours}^{-1}. \tag{11.36}$$

Substituting (11.36) in (11.35)

$$(\text{MTBF})_s = \frac{2}{1/7083} = 2 \times 7083 = 14,166\ \text{hours},$$

which is equal to one year, plus 32 weeks, plus one day, plus six hours (less than half the MTBF of a single drive!).

It would be very easy to think that the inclusion of a "hot spare" is equivalent to having another active redundant drive. However, this is *not* so. It is very easy to show this mathematically by considering two active redundant drives, that is,

$$k = 8, \quad n = 10.$$

Substituting those values in Eq. (11.19)

$$(\text{MTBF})_s = \frac{1}{\lambda} \left[\frac{1}{8} + \frac{1}{9} + \frac{1}{10} \right] = \frac{1}{\lambda} \left[\frac{121}{360} \right]. \tag{11.37}$$

Comparing equations (11.20), (11.35), and (11.37)

$$\frac{17}{32} < \frac{121}{360} \ll 2.$$

That is

$$0.236 < 0.336 \ll 2. \tag{11.38}$$

Equation (11.38) can be interpreted by stating that adding another active redundant drive increases the reliability somewhat (after all, $0.336 > 0.236$), although not by much. However, if another drive is added *not* as another active redundant drive but as a "hot spare," the reliability is increased considerably (after all, $2 \gg 0.336$). Consequently, the two methods of redundancy are completely different.

Case 2. Reconstruction of Failed Drive. We calculated already the MTBF for several reconstruction times. Some of these values were shown in Table 11.1. and plotted in Fig. 11.8. Therefore, according to Eq. (11.35) it is only necessary to double those figures. That is, according to eq. (11.35) and (11.29),

$$(\text{MTBF})_s \cong 2 \times \frac{25 \times 10^6}{T} = \frac{50 \times 10^6}{T} \text{ hours}, \tag{11.39}$$

which increases the reliability substantially. Therefore, the standby redundancy should be adopted for the case of disk arrays, simply because it constitutes a rather inexpensive way of increasing the reliability substantially.

In conclusion, the previous calculations point towards an optimum disk array configuration for this particular case of one byte in parallel (eight bits):

(1) Use 10 drives (one as active redundancy, the other as hot spare).
(2) Make use of the ECC algorithms already employed extensively in nine-track high-density tape drives.

(3) When one of the disk drives fails and becomes non-operational, the ECC mentioned above should be able to produce (on the fly) a good byte, by means of the remaining eight drives.
(4) Remove the failed drive. Switch in the hot spare and reconstruct the data previously contained in the failed drive. When this operation is finished we should have again nine good drives. The reconstruction time should be nearly 15 minutes.
(5) Install a new disk drive in the place previously occupied by the failed drive. This new drive now becomes the hot spare.
(6) Alghough the hot spare drive is not used during normal operation, all the drives in the system (namely, 10) should be synchronized at all times. This should facilitate (and diminish) the reconstruction, when needed.

The examples shown above are only indicative, not exhaustive, and they were illustrated mainly to show a methodology. However, most of the formulas are general, and, within the constraints imposed by the assumptions, they should be applicable to many other cases and limited only by the imagination of the designer or user.

11.8. SUMMARY

This chapter has presented a method whereby the data rate and reliability have been increased considerably. This example has served to indicate the results that can be achieved at the system level (above and beyond the characteristics of a single DASD). Still, the slow access to data must be resolved. This has been attempted in the past by techniques such as "shadowing," "interleaving," etc. At this point in time, a great deal remains to be done by the computer scientists to overcome this important limitation, considering the high data rates and reliabilities demanded by present-day data processing equipment.

Bibliography

Adler, R. L., Hassner, M. and Moussouris, P. "Method and apparatus for generating a noiseless sliding block code for a $(1, 7)$ channel with rate 2/3." U.S. Patent No. 4,413,251 (1982).

Adler, R., Coppersmith, D. and Hassner, M. "Algorithms for sliding block codes." *IEEE Transactions on Information Theory* **IT-29–1** (January 1983).

Armitage, J. D. and Cannon, M. R. "Magnetic recording and read-back systems with raised-cosine equalization." U.S. Patent No. 3,775,759 (November 1973).

Balabanian, N. "Network Synthesis." Prentice-Hall, Inc. (1958).

Barlow, R. E. and Proscham, F. "Mathematical Theory of Reliability." John Wiley and Sons (1965).

Bateman, H. "Higher transcendental function, Vols. 1 and 2." McGraw-Hill Book Co. (1953).

Bayer, R. G. "A theoretical treatment of self-demagnetization in magnetic recording." *IEEE Transactions on Audio* **AU-11,** pp. 81–88 (May/June 1963).

Bazovsky, I. "Reliability theory and practice." Prentice-Hall, Englewood Cliffs, New Jersey (1961).

Beckenhauer, R. L. and Schaeuble, W. J. "Phase alignment of 1F and 2F clocks." *IBM Technical Disclosure Bulletin* **21–1,** pp. 326–327 (June 1978).

Beckenhauer, R. L. and Schaeuble, W. J. "Sync pattern encoding scheme for Run-Length-Limited codes." U.S. Patent No. 4,146,909 (March 1979).

Becker, F. H., Kretzmer, E. R. and Sheehan, J. R. "A new signal format for efficient data transmission." *The Bell System Technical Journal,* pp. 755–758 (May/June 1966).

Behr, M. I. and Blessum, N. S. "Technique for reducing effects of pulse crowding in magnetic recording." *IEEE Transactions on Magnetics* **MAG-8-5,** pp. 608–611 (September 1972).

Bell, C. G. "The mini and micro industries." *IEEE Transactions on Computers* **17–10,** pp. 14–30 (October 1984).

Bonyhard, P. I., Middleton, B. K. and Davies, A. V. "The dependence of maximum recording density on the properties of the recording medium." *IEEE Transactions on Magnetics* (correspondence section) **MAG-1,** pp. 423–424 (December 1965).

Bonyhard, P. I., Davies, A. V. and Middleton, B. K. "A theory of digital magnetic recording on metallic films." *IEEE Transactions on Magnetics,* **MAG-2,** pp. 1–5 (March 1966).

Bowden, F. P. and Tabor, D. "Friction: An Introduction to Tribology." Anchor Press/Doubleday (1973).

Bracewell, R. M. "The Fourier transform and its properties." McGraw-Hill Book Co. (1965).

Buchanan, J. E. "Build a high-frequency phase-locked loop with common IC components." *Electronic Design* **16,** p. 108 (August 1977).

Burkhardt, H. "Phase detection with run-length limited codes." *IBM Technical Disclosure Bulletin* **24–1B,** p. 683 (June 1981).

Burkhardt, H. "An event-driven maximum-likelihood peak position detector for run-length-limited codes in Magnetic Recording." IBM Research Report RJ3167 (June 24, 1981).

Calabro, S. R. "Reliability Principles and Practices." McGraw-Hill Book Co. (1962).

Camras, M. "Magnetic Recording Handbook." Van Nostrand Reinhold (1988).

Chao, S. C. "Binary code magnetic recording system." U.S. Patent No. 3,277,454 (October 1966).

Chen, C. "Active filter design." Hayden Book Co. (1982).

Chi, C. S. "Characterization and spectral equalization for high-density disk recording." *IEEE Transactions on Magnetics* **MAG-15-6,** pp. 1447–1449 (November 1979).

Chi, C. S. "Spacing loss and nonlinear distortion in digital magnetic recording." *IEEE Transactions on Magnetics* **MAG-16-5,** pp. 976–978 (September 1980).

Cohn, M., Jacoby, G. V. and Bates III, A. "Data encoding method and system employing two-thirds code rate with full word look-ahead." U.S. Patent No. 4,337,458 (1982).

Cohn, M. and Jacoby, G. V. "Run length reduction of 3PM code via look-ahead techniques." *IEEE Transactions on Magnetics* **MAG-18-6** (November 1982).

Cullum, C. D. "Encoding and signal processing." *Annals of the New York Academy of Sciences* **189** pp. 52–62 (January 3, 1972).

Curland, N. and Speliotis, D. E. "A theoretical study of an isolated transition using an iterative hysteretic model." *IEEE Transactions on Magnetics* **MAG-6,** pp. 640–646 (September 1970).

Dhillon, B. S. and Singh, C. "Engineering reliability—new techniques and applications." John Wiley and Sons (1981).

Disk Storage Technology. IBM Document No. GA-26-1665-0 (1980).

Dunn, G. A. "Phase setting for a (2, 7) code decoder." *IBM Technical Disclosure Bulletin* **23–4,** pp. 1673–1675 (September 1980).

Eggenberger, J. S. and Hodges, P. "Sequential encoding and decoding of variable-word-length, fixed rate data codes." U.S. Patent No. 4,115,768 (September 19, 1978).

Eldridge, D. F. "Magnetic recording and reproduction of pulses." *IRE Transactions on Audio* **AU-9,** pp. 42–57 (April 1960).

Fan, G. J. "A study of the playback process of a magnetic ring head." *IBM Journal of Research and Development* **5,** p. 321 (October 1961).

Fan, G. J. and Tang, D. T. "An optimal time-domain filter for recording channels." IBM Research Report RC-1328 (December 9, 1964).

Feth, G. O. "Analysis of magnetic recording fields." *AIEE Transactions (Commun. Electron.)* **81,** pp. 267–279 (September 1962).

Fisher, R. D. and Newman, J. J. "Code performance and head/media interface." *IEEE Transactions on Magnetics* **MAG-17–4,** pp. 1452–1454 (July 1981).

Fisher, R. D. and Taranto, J. "Vertical recording brings disk drives standing-room-only densities." *Electronics,* pp. 131–135 (September 22, 1982).

Franaszek, P. A. "Sequence-state coding for digital transmission." *The Bell System Technical Journal,* pp. 143–157 (November 1967).

Franaszek, P. A. "Determination of the existence of codes for synchronous transmission of binary data over discrete constrained channels." IBM Research Report RC 2468 (October 3, 1968).

Franaszek, P. A. "Sequence-state methods for run-length-limited coding." *IBM Journal of Research and Development* **14–4,** pp. 376–383 (July 1970).

Franaszek, P. A. "Run-length-limited variable-length coding with error propagation limitation." U.S. Patent No. 3,689,899 (September 1972).

Franaszek, P. A. "Efficient code for digital magnetic recording." *IBM Technical Disclosure Bulletin* **23-9** (February 1981).

Franchini, R. C. and Warner, D. L. "A method of high-density recording on flexible magnetic disks." *Computer Design,* pp. 106–109. (October 1976)

Gabor, A. "High-density recording on magnetic tape." *Electronics* **32-42,** pp. 72–75 (October 16, 1959).

Gabor, A. "Adaptive coding for self-clocking recording." *IEEE Transactions on Electronic Computers* **EC-16-6,** pp. 866–868 (December 1967).

Gardner, M. F. and Barnes, J. L. "Transients in linear systems." John Wiley and Sons, Inc. (1956).

Gardner, F. M. "Phaselock techniques." John Wiley and Sons (1966).

Geffon, A. P. "A 6 KBPI storage system using Mod-11 interface." *IEEE Transactions on Magnetics* **MAG-13-5,** pp. 1205–1207 (September 1977).

George, D. I., King, S. F. and Carr, A. E. "A self-consistent calculation of the magnetic transition recorded on a thin film disk." *IEEE Transactions on Magnetics* **MAG-7,** pp. 240–243 (June 1971).

Glover, N. "Practical error correction design for engineers." Data Systems Technology Corp., Broomfield, Colorado (1982).

Glover, N. "Data accuracy issues for disk controller developers." *Computer Design,* p. 127 (October 1982).

Goretski, J. A., *et al.* "Detector for eliminating readback errors from disk files." *IBM Technical Disclosure Bulletin* **23-10,** pp. 4639–4641 (March 1981).

Gorog, E. "A new approach to time-domain equalization." *IBM Journal of Research and Development* **9-4,** pp. 228–232 (July 1965).

Gorog, E. "Redundant alphabets with desirable frequency spectrum properties." *IBM Journal of Research and Development* **12-3** (May 1968).

Graham, I. H. *et al.* "Signal detection circuit." U.S. Patent No. 3,631,263 (December 1971).

Grebene, A. B. "A monolithic phase-locked signal conditioner/demodulator (NE 560/561)." Signetics Corp. (1970).

Hamming, R. W. "Coding and information theory." Prentice-Hall, Englewood Cliffs, New Jersey (1980).

Harker, J. M. *et al.* "A quarter century of disk file innovation." *IBM Journal of Research and Development* **25-5,** pp. 677–689 (September 1981).

Harker, J. "Byte oriented data compression techniques." *Computer Design,* pp. 95–100 (October 1982).

Haynes, M. K. "Experimental determination of the loss and phase transfer functions of a magnetic recording channel." *IEEE Transactions on Magnetics* **MAG-13,** pp. 1284–1286 (September 1977).

Heidecker, R. F. "Predifferentiated recording." U.S. Patent No. 3,603,942 (September 1971).

Herbert, J. R. and Patterson, D. W. "A computer simulation of the magnetic recording process." *IEEE Transactions on Magnetics* **MAG-1,** pp. 352–357 (December 1965).

Hilburn, J. L. and Johnson, D. E. "Manual of active filter design." McGraw-Hill Book Co. (1973).

Hodges, P. "Encoding and decoding checking arrangement for 2-7 run-length limited codes." *IBM Technical Disclosure Bulletin* **23–4,** pp. 1635–1636 (September 1980).

Hoepner, J. F. and Wall, L. H. "Encoding/decoding techniques double floppy disk capacity." *Computer Design,* pp. 127–135 (February 1980).

Horiguchi, T. and Morita, K. "On optimization of modulation codes in digital recording." *IEEE Transactions on Magnetics* **MAG-12–6,** pp. 740–742 (November 1976).

Huber, W. D. "Maximization of linear recording density." *IEEE Transactions on Magnetics* **MAG-13–5,** pp. 1208–1210 (September 1977).

Huber, W. D. "Selection of modulation code parameters for maximum linear density." *IEEE Transactions on Magnetics* **MAG-16–5,** pp. 637–639 (September 1980).

Huelsman, L. P. "Theory and design of active RC circuits." McGraw-Hill Book Co. (1968).

Hughes, G. F. and Schmidt, R. K. "On noise in digital recording." *IEEE Transactions on Magnetics* **MAG-12–6,** pp. 752–754 (November 1976).

Hung, J. W. "Transfer function and error probability of a digital tape recording system." *Journal of Applied Physics,* **31–5,** pp. 396 S and 397 S (May 1960).

IBM 3380 Direct Access Storage Introduction. IBM GC 26-4491-0 (September 1987).

Ireson, W. G. "Reliability handbook." McGraw-Hill Book Co. (1966).

Iwasaki, S. I. and Suzuki, T. "Dynamical interpretation of magnetic recording process." *IEEE Transactions on Magnetics* **MAG-4,** pp. 269–276 (September 1968).

Iwasaki, S. I. and Nakamura, Y. "An analysis of magnetization mode for high density magnetic recording." *IEEE Transactions on Magnetics* **MAG-13–5,** pp. 1272–1277 (September 1977).

Iwasaki, S. I. and Ouchi, K. "CoCr recording films with perpendicular magnetic anisotropy." *IEEE Transactions on Magnetics* **MAG-14–5,** pp. 849–851 (September 1978).

Iwasaki, S. I., Nakamura, Y. and Ouchi, K. "Perpendicular magnetic recording with a composite anisotropy medium." *IEEE Transactions on Magnetics,* **MAG-15,** pp. 1456–1458 (September 1979).

Jackson, D. "Fourier series and orthogonal polynomials." The Mathematical Association of America (1941).

Jacoby, G. V. "Signal equalization in digital magnetic recording." *IEEE Transactions on Magnetics* **MAG-11–5,** pp. 302–305 (September 1975).

Jacoby, G. V. "A New look-ahead code for increased data density." *IEEE Transactions on Magnetics* **MAG-13–5,** pp. 1202–1204 (September 1977).

Jacoby, G. V. and Kost, R. "Binary two-thirds rate code with full word look-ahead." *IEEE Transactions on Magnetics* **MAG-20–5** (September 1984).

Javid, B. "New techniques squeeze more data onto magnetic tape." EDN, pp. 63–64 (June 5, 1975).

Jenkins, R. E. "1F/2F phase alignment system." *IBM Technical Disclosure Bulletin* **23–1,** pp. 318–319 (June 1980).

Jensen, L. A. "Electronic remote supervision and control of states and processes." U.S. Patent No. 3,689,890 (April 1970).

Jilke, W. "Disk array mass storage systems: The new opportunity." Amperiff Corp. (September 30, 1986).

Johnson, D. E., Johnson, J. R. and Moore, H. P. "A handbook of active filters." Prentice-Hall, Inc., Englewood Cliffs, New Jersey (1980).

Jorgensen, F. "The complete handbook of magnetic recording." TAB Books, No. 1059 (1980).

Kalstrom, D. J. "Simple encoding schemes double capacity of a flexible disk." *Computer Design,* pp. 98–102 (September 1976).

Kameyama, T., *et al.* "Improvement of recording density by means of cosine equalizer." *IEEE Transactions on Magnetics* **MAG-12–6,** pp. 746–748 (November 1976).

Katz, E. R. and Campbell, T. G. "Effect of bitshift distribution on error rate in digital magnetic recording." *IEEE Transactions on Magnetics* **MAG-15–3,** pp. 1050–1053 (May 1979).

Kautz, W. H. "Network synthesis for specified transient response." MIT Research Laboratory of Electronics Technical Report No. 209 (April 23, 1952).

Kautz, W. H. "Fibonacci codes for synchronization control." *IEEE Transactions on Information Theory* **IT-11,** pp. 284–292 (April 1965).

Kiwimagi, R. G., *et al.* "Channel coding for digital recording." *IEEE Transactions on Magnetics* **MAG-10–3,** pp. 515–518 (September 1974).

Kobayashi, H. "On the relationship between the channel frequency characteristic and intersymbol interference." IBM Research Report RC 2129 (April 1968).

Kobayashi, H. "A simulation study of the performance of correlative level coding schemes and special signalling alphabets." IBM Research Report RC 2229 (October 1968).

Kobayashi, H. "Coding schemes for reduction of intersymbol interference in data transmission systems." *IBM Journal of Research and Development*, pp. 343–353 (July 1970).

Kobayashi, H. and Tang, D. T. "Application of partial-response channel coding to magnetic recording." *IBM Journal of Research and Development*, pp. 368–375 (July 1970).

Koshtyshyn, B. "A harmonic analysis of saturation recording in a magnetic medium." *IRE Transactions on Electronic Computers* **EC-11,** pp. 253–263 (April 1962).

Koshtyshyn, B. "A theoretical model for a quantitative evaluation of magnetic recording systems." *IEEE Transactions on Magnetics* **MAG-2,** pp. 236–242 (September 1966).

Kretzmer, E. R. "Generalization of a technique for binary data communication." *IEEE Transactions on Communication Technology* **COM-14,** pp. 67–68 (February 1966).

Kuh, E. S. and Pederson, D. O. "Principles of circuit synthesis." McGraw-Hill Book Co. (1959).

Kurzweil, F. "A universal intelligent micro-controller-based spindle drive system for hard disk drives." Paper presented at the conference on Applied Motion Control, University of Minneapolis, Minnesota, June 16 to 18, 1987.

Kurzweil, F. "Challenges in Winchester Technology—disk drive arrays." Symposium at the University of Santa Clara, California (December 17, 1987).

Labinger, R. L. "Use an N-bit detector for phase-locking." *Electronic Design*, **20,** pp. 44–47 (September 30, 1971).

Lancaster, D. "Active filter cookbook." Howard W. Sams & Co., No. 21,168, (1975).

Langland, B. J. and Larimore, M. G. "Processing of signals from media with perpendicular anisotropy." *IEEE Transactions on Magnetics* **MAG-16-5,** pp. 640–642 (September 1980).

Langland, B. T. and Albert, P. A. "Recording on perpendicular anisotropy media with ring heads." *IEEE Transactions on Magnetics* **MAG-17,** pp. 2547–2549 (1981).

Lempel, A. and Cohn, M. "Look-ahead coding for input restricted channels." *IEEE Transactions on Information Theory* **IT-28-6** (November 1982).

Lender, A. "Correlative level coding for binary-data transmission." *IEEE Spectrum,* pp. 104–115 (February 1966).

Lloyd, D. K. and Lipow, M. "Reliability: Management, methods and mathematics." Prentice-Hall, Englewood Cliffs, New Jersey, Second Edition (1977).

Lowman, C. E. "Magnetic recording." McGraw-Hill Book Co. (1972).

Lucky, L. W., Salz, J. S. and Weldon, E. J. "Principles of data communication." McGraw-Hill Book Co. (1968).

Mackintosh, N. D. "The choice of a recording code." *IERE Conference Proceedings,* paper No. 43, pp. 77–119 (July 24 to 27, 1979).

Mackintosh, N. D., "A superposition-based analysis of pulse-slimming techniques for digital recording." Video and Data Recording Conference (G.B.) (1979). Also: *Radio and Electronic Engineer* **50–6,** pp. 307–314 (June 1980).

Mackintosh, N. D. "The choice of a recording code." *Institution of Electronic and Radio Engineers* (G.B.) **50–4,** pp. 117–193 (April 1980).

Mackintosh, N. D. "Evaluation of disk drives by margin analysis." Presented at the International Peripherals and Software Exposition, Anaheim Convention Center, Anaheim, California (October 1, 1982).

Maginnis, N. B. "Store more, spend less: Mid-range options abound." *Computer-world* p. 71 (November 16, 1987).

Magnus, W. and Oberhettinger, F. "Formulas and theorems for the functions of mathematical physics." Chelsea Publishing Co. (1954).

Mallinson, J. C. "Demagnetization theory for longitudinal recording." *IEEE Transactions on Magnetics* **MAG-2** pp. 233–235 (September 1966). Correction on **MAG-3,** p. 171 (June 1967).

Mallinson, J. C. "Novel technique for the measurement of demagnetizing fields in longitudinal recording.' *IEEE Transactions on Magnetics* **MAG-3,** pp. 201–204 (September 1967).

Mallinson, J. C. "A theoretical limit to digital pulse resolution in saturation recording." *IEEE Transactions on Magnetics* **MAG-5–2,** pp. 91 –97 (June 1969).

Mallinson, J. C. "Maximum signal-to-noise ratio of a tape recorder." *IEEE Transactions on Magnetics* **MAG-5–3,** pp. 182–186 (September 1969).

Mallinson, J. C. and Miller, J. W. "On optimal codes for digital magnetic recording." Proceedings of the 1976 Birmingham Conference on Video and Digital Recording, pp. 161–169. Also: *Radio and Electronic Engineer.* **47–4,** pp. 172–176 (April 1977).

Massey, W. C. "A two-seven designer's guide for digital magnetic recording" (1982). William C. Massey, Broomfield, Colorado (303)466-5923.

Mattis, J. A. and Camenzind, H. R. "A new phase locked loop with high stability and accuracy (SE/NE 565)." Signetics Corp. Document No. D 147-1 LIN-018-110 10M (1971).

McClellan, J. M., Parks, T. W. and Rabiner, L. R. "A computer program for designing optimum FIR linear phase filters." *IEEE Transactions on Audio and Electroacoustics* **21**, p. 506 (1973).

McLeod, J. "Winchester Disks in microcomputers." Elsevier Science Publishers b.v. Amsterdam/Information and Business Division (1983).

Melas, C. M. "Quelques proprietes des codes binaires avec symbols consecutifs." IEEE International Symposium on Information Theory (1967).

Melas, C. M. and Corog, E. "A digital data transmission system with optimum bandwidth utilization." Proceedings of the National Communications Conference, Utica, New York, pp. 330–341 (October 1964).

Middleton, B. K. "The dependence of recording characteristics on thin metal tapes on their magnetic properties and on the replay head." *IEEE Transactions on Magnetics* **MAG-2,** pp. 225–229 (September 1966).

Miessler, M. "Translator for run-length-limited code." *IBM Technical Disclosure Bulletin* **17–5,** pp. 1489–1491 (October 1974).

Miller, A. U.S. Patent No. 3,108,261 (October 22, 1963).

Miller, J. W. U.S. Patent No. 4,027,335 (May 31, 1977).

Minukhin, V. B. "Phase distorsions of signals in magnetic recording equipment." *Telecommunications Radio Engineering* **29–30,** pp. 114–120 (1975).

Miyata, J. J. and Hartel, R. R. "The recording and reproduction of signals on magnetic medium using saturation type recording." *IRE Transactions on Electronic Computer* **EC-8,** pp. 159–169 (June 1959).

Mokhoff, N. "Parallel disk assembly packs 1.5 G Bytes, runs at 4 M Bytes/s. *Electronic Design,* pp. 45–46 (November 12, 1987).

Mommes, J. H. and Raviv, J. "Coding for data compaction." IBM Research Report No. RC 5150 (November 26, 1974).

Monett, M. R. "Data recovery circuit." U.S. Patent No. 3,810,234 (May 7, 1974).

Morgan, D. P. "Surface-wave devices for signal processing." Elsevier Science Publishers B.V. (1985).

Morriss, D. J. "Code your fiber-optic data for speed, without losing circuit simplicity." *Electronic Design* **22,** pp. 84–91 (October 25, 1978).

Moschytz, G. S. "Miniaturized RC filters using phase-locked loops." *Bell System Technical Journal* **44,** pp. 823–870 (May 1965).

Myers, W. "The competitiveness of the United States disk industry." *IEEE Transactions on Computers* **19–11,** pp. 85–90 (January 1986).

Newman, J. J., and Fisher, R. D. "Signal-to-noise limitations of particulate disk coatings." *IEEE Transactions on Magnetics* **MAG-16–1,** pp. 23–25 (January 1980).

Norton, J. J. "Drop your costs, but not your bits with a Manchester-data decoder." *Electronic Design,* pp. 110–116 (July 19, 1979).

O'Connor, P. D. T. "Practical reliability engineering." John Wiley and Sons, Second Edition (1985).

Oswald, R. K. "Design of a disk file head-positioning servo." *IBM Journal of Research and Development* **18–6,** pp. 506–512 (November 1974).

Oswald, R. K. "An electronic tachometer for disk file motion control." *IEEE Transactions on Magnetics* **MAG-11–5,** pp. 1245–1246 (September 1975).

Oswald, R. K. "Head positioning servo for the IBM 3344/3350 disk files." *IEEE Transactions on Magnetics* **MAG-14–4,** pp. 176–177 (July 1978).

Oswald, R. K. "The IBM 3370 head-positioning control system." IBM Disk Storage Technology, No. GA26-1665, pp. 41–44 (February 1980).

Park, A. and Balasubramanian, K. "Providing fault tolerance in parallel secondary storage systems." Department of Computer Science, Princeton University, C.S.-TR-057-86 (November 7, 1986).

Pasternak, G. and Whalin, R. L. "Analysis and synthesis of a digital phase-locked loop for FM demodulation." *Bell System Technical Journal* **47,** pp. 2207–2237 (December 1968).

Patel, A. M. "Zero-modulation encoding in magnetic recording." *IBM Journal of Research and Development* **19–4,** pp. 366–378 (July 1975). Also: U.S. Patent No. 3,810,111.

Patel, A. M. and Hong, S. J. "Optimal rectangular code for high density magnetic tapes." *IBM Journal of Research and Development,* pp. 579–587 (November 1974).

Patterson, D. A., Gibson, G. and Katz, R. H. "A case for redundant arrays of inexpensive disks (RAID)." Computer Science Division, Department of Electrical Engineering and Computer Sciences, University of California, Berkeley (December 1987).

Pau, L. F. "Failure diagnosis and performance monitoring." Marcel Dekker, Inc., New York (1975).

Perron, O. "Die lehre von den kettenbruchen." Chelsea Publishing Co. (1950).

Phelps, B. E. "Magnetic recording method." U.S. Patent No. 2,774,646 (1956).

Porter, J. N. 1987 Disk/trend report—rigid disk drives. Disk/Trend, Inc., Mountain View, California, (415)961-6209.

Potter, R. I. "Analysis of saturation magnetic recording based on arctangent magnetization transitions." *Journal of Applied Physics* **41**, pp. 1647–1651 (March 1970).

Potter, R. I. and Beardsley, I. A. "Self-consistent computer calculations for perpendicular magnetic recording." Paper 23.3, INTERMAG (1980). Also: *IEEE Transactions on Magnetics* **MAG-16–5**, pp. 967–972 (September 1980).

Price, R., *et al.* "Dual channel signal detector circuit." U.S. Patent No. 4,081,756 (March 1978).

Reghbati, H. K. "An overview of data compression techniques." *Computer* pp. 71–75 (April 1981).

Ringkjøb, E. T. "Achieving a fast data-transfer rate by optimizing existing technology." *Electronics,* pp. 86–91 (May 1, 1975).

Rodman, D. "Line coding techniques for digital transmission." *Electronic Products,* pp. 71–76 (September 30, 1982).

Sanders, L. "Improve datacommunication links by using Manchester code." *EDN,* pp. 155–162 (February 17, 1982).

Schneider, R. C. "Modified double frequency encoding." *IBM Technical Disclosure Bulletin* **8–11,** p. 1475 (April 1966).

Schneider, R. C. "Phase shift reducing digital signal recording having no DC component." U.S. Patent No. 3,855,616 (December 1974).

Schneider, R. C. "An improved pulse-slimming method for magnetic recording." *IEEE Transactions on Magnetics* **MAG-11–5,** pp. 1240–1241 (September 1975).

Schneider, R. C. "Write equalization in high-linear-density magnetic recording." *IBM Journal of Research and Development* **29–6,** pp. 563–568 (November 1985).

Schwarz, T. A. "A statistical model for determining the error rate of the recording channel." *IEEE Transactions on Magnetics* **MAG-16–5,** pp. 634–636 (September 1980).

Sharpe, C. A. "A three-state phase detector can improve your next PLL design." *EDN,* pp. 55–59 (September 20, 1976).

Sherer, P. "Anti-shouldering read circuit for disk magnetic memory." U.S. Patent No. 4,016,599 (April 1977).

Shooman, M. L. "Probabilistic reliability: An engineering approach." McGraw-Hill Book Co. (1968).

Sidhu, P. S. "Group-coded recording reliably doubles diskette capacity." *Computer Design,* pp. 84–88 (December 1976).

Siegel, P. H. "Applications of a peak detection channel model." *IEEE Transactions on Magnetics,* **MAG-18–6** (November 1982).

Siegel, P. H. "Recording codes for digital magnetic storage." *IEEE Transactions on Magnetic* **MAG-21–5,** pp. 1344–1349 (September 1985).

Sierra, H. M. "Increased digital magnetic recording readback resolution by means of a linear, passive network." *IBM Journal of Research and Development* **7–1,** pp. 22–33 (January 1963). Also: U.S. Patent No. 3,215,995 (November 2, 1965).

Sierra, H. M. "Extended studies on linear filtering of magnetic recording read-back pulses." IBM Technical Report TR-02.244 (November 30, 1962).

Sierra, H. M. "Design of a pulse-narrowing network." *Electrotechnology,* pp. 38–42 (September 1963).

Sierra, H. M. "White noise, linear, passive, Gaussian pulse, matched filter." *IBM Technical Disclosure Bulletin,* **6–6,** pp. 52–53 (November 1963).

Sierra, H. M. "Linear, passive, matched filter for digital magnetic recording." *IEEE Transactions on Electronic Computers* **EC-14–2,** pp. 204–209 (April 1965).

Sierra, H. M. "Estimate of pulse width in digital magnetic recording." *IEEE Proceedings* **53–5,** pp. 513–514 (May 1965).

Sierra, H. M. "Bit shift and crowding in digital magnetic recording." *Electro-technology,* pp. 56-58 (September 1966).

Sierra, H. M. "Optimum surface partitioning in random access disk files." *IBM Technical Disclosure Bulletin* **12–12,** pp. 2366–2368 (May 1970).

Sierra, H. M. "Information storage and retrieval by electron optics." *IBM Technical Disclosure Bulletin* **13–6,** pp. 1638–1640 (November 1970).

Sierra, H. M. "Digital magnetic recording read channel using statistical decisions for data detection." *IBM Technical Disclosure Bulletin* **15–2,** p. 514 (July 1972).

Sierra, H. M. "Active RC filters without inductors." *IBM Technical Disclosure Bulletin* **15–6,** p. 1788 (November 1972).

Sierra, H. M. "DC restoration by a constant-R symmetrical, passive lattice." *IBM Technical Disclosure Bulletin* **16–12,** pp. 4049–4050 (May 1974).

Sierra, H. M. "Wideband active filter." *IBM Technical Disclosure Bulletin* **17–2,** pp. 533–535 (July 1974).

Sierra, H. M. "Read channel test circuit." *IBM Technical Disclosure Bulletin* **19–5,** pp. 1911–1912 (October 1976).

Sierra, H. M. "Disk Storage Trends: Past, Present & Future." Insights (October 1988).

Sierra, H. M. "The Astounding Promise of Disk Arrays." Insights (Winter 1989). Also: Canadian Data Systems **21–5,** p. 52 (May 1989).

Slater, L. J. "Confluent hypergeometric functions." Cambridge University Press (1960).

Speliotis, D. E. "Magnetic recording theories: Accomplishments and unresolved problems." *IEEE Transactions on Magnetics* **MAG-3,** pp. 195–200 (September 1967).

Speliotis, D. E. and Morrison, J. R. "A theoretical analysis of saturation magnetic recording." *IBM Journal of Research and Development* **10,** pp. 233–243 (May 1966).

Spitzer, C. F. "Digital magnetic recording of wideband analog signals." *Computer Design,* pp. 83–90 (September 1973).

Steele, C. W. "A computer simulation of unbiased digital recording." *IEEE Transactions on Magnetics* **MAG-4–4** (December 1968).

Stevens, L. D. "The evolution of magnetic storage." *IBM Journal of Research and Development* **25–5,** pp. 663–675 (September 1981).

Stevenson, T. J. "Disk file optimization." *IEEE Transactions on Magnetics* **MAG-11–5,** pp. 1237–1239 (September 1975).

Stieltjes, T. J. "Recherches sur les fractions continues." *Ann. Fac. Sci. Toulouse* **8,** pp. 1–122 (1894).

Su, J. L. and Williams, M. L. "Noise in disk data-recording media." *IBM Journal of Research and Development,* pp. 570–575 (November 1974).

Tachibana, M., *et al.* "Optimum waveform design and its effect on the peak shift compensation." *IEEE Transactions on Magnetics* **MAG-13–5,** pp. 1199–1201 (September 1977).

Tahara, Y., *et al.* "Optimum design of channel filters for digital magnetic recording." *IEEE Transactions on Magnetics* **MAG-12–6,** pp. 740–742 (November 1976).

Tahara, Y., *et al.* "Peak shift caused by Gaussian Noise in digital magnetic recording." *Electronics and Communications in Japan* **59-C–10,** pp. 77–86 (1976).

Talbot, A. "A new method of synthesis of reactance networks." *Proceedings of the I.E.E.* (London), Part IV, No. 101, Monograph No. 77, pp. 73–90.

Tamura, T., *et al.* "A coding method in digital magnetic recording." *IEEE Transactions on Magnetics* **MAG-8–6,** pp. 612–614 (September 1972).

Tang, D. T. "Run-length limited codes for synchronization and compaction." IBM Research Report RC1883 (August 1, 1967).

Tang, D. T. "Practical coding schemes with run-length constraints." IBM Research Report No. RC-2022 (April 25, 1968).

Tang, D. T. "Run-length-limited codes." IEEE International Symposium on Information Theory (1969).

Tang, D. T. "Run-length-limited coding for modified raised-cosine equalization channel." U.S. Patent No. 3,647,964 (March 1972).

Tang, Y-S. "Noise autocorrelation in magnetic recording systems." *IEEE*

Transactions on Magnetics **MAG-21-5,** pp. 1389–1391 (September 1985).

Truxal, J. G. "Numerical analysis for network design." *I.R.E. Transaction on circuit Theory,* pp. 49–60 (September 1954).

Tufts, D. W. "Nyquist's problem—The joint optimization of transmitter and receiver in pulse amplitude modulation." *Proceedings of the IEEE* **53** (March 1965).

Tuttle, D. F. "Network synthesis, Vol. 1." John Wiley and Sons, Inc. (1958).

Van Valkenburg, M. E. "Introduction to modern network synthesis." John wiley and Sons, Inc. (1960).

Veillard, D. H. "Compact spectrum recording, a new binary process maximizing the use of a recording channel." *IEEE Transactions on Magnetics* **MAG-20-5,** pp. 891–893 (September 1984).

Viterbi, A. J. "Principles of coherent communication." McGraw-Hill Book Co. (1966).

Wall, H. S. "Analytic theory of continued fractions." D. Van Nostrand Co. (1948).

Wallace, R. L. "The reproduction of magnetically recorded signals." *Bell System Technical Journal* **30,** pp. 1146–1173 (October 1951).

Warner, M. U.S. Patent No. 3,823,416 (July 9, 1974).

Warren, C. "Vertical, quasivertical recording schemes promise new highs in disk density." *Electronic Design,* pp. 43–44 (February 3, 1983).

Webster's Ninth New Collegiate Dictionary. Merriam-Webster Inc., Publishers (1987).

Weldon, E. J. "Performance of a forward-acting error-control system on the switched telephone network." *The Bell System Technical Journal,* pp. 756–764 (May/June 1966).

Westmijze, W. K. "Studies in magnetic recording." *Philips Research Reports* **8** pp. 245–269 (June 1953).

Wynn, P. "The rational approximation of functions which are formally defined by a power series expansion." *Mathematics of Computation,* pp. 147–186 (April 1960).

Yencharis, L. "Small Winchester drives move up to mainframe encoding schemes." *Electronic Design,* pp. 51–52 (October 15, 1981).

Zverev, A. I. "Handbook of filter synthesis." John Wiley and Sons, Inc. (1967).

Manufacturers of Direct Access Storage Devices

U. S. MANUFACTURERS

ALPHA DATA, INC.
20750 Marilla Street
Chatsworth, CA 91311

AMPEX CORPORATION
401 Broadway
Redwood City, CA 94063

BRAND TECHNOLOGIES, INC.
6140 Variel Avenue
Woodland Hills, CA

CARDIFF PERIPHERALS CORPORATION
5421 Avenida Encinas
Carlsbad, CA 92008

CENTURY DATA, INC.
2055 Gateway Place
San Jose, CA 95110

CONNER PERIPHERALS, INC.
2221 Old Oakland Road
San Jose, CA 95131

CONTROL DATA CORPORATION
8100 34th Avenue South
Minneapolis, MN 55440

DATA GENERAL CORPORATION
4400 Computer Drive
Westboro, MA 01581

DATA-TECH MEMORIES, INC.
2350 Shasta Way
Simi Valley, CA 93065

DIGITAL EQUIPMENT CORPORATION
146 Main Street
Maynard, MA 01754

DISC TECH ONE
849 Ward Drive
Santa Barbara, CA 93111

DMA TECHNOLOGIES
601 Pine Avenue
Goleta, CA 93117

HEWLETT–PACKARD COMPANY
3000 Hanover Street
Palo Alto, CA 94303

IBIS SYSTEMS, INC.
5775 North Lindero Canyon Drive
Westlake Village, CA 91360

INTERNATIONAL BUSINESS MACHINES CORPORATION
Route 22
Armonk, NY 10504

JOSEPHINE COUNTY TECHNOLOGY, INC.
1899 N. W. Hawthorne
Grants Pass, OR 97526

MAXTOR CORPORATION
150 River Oaks Parkway
San Jose, CA 95134

MEMOREX CORPORATION
611 South Milpitas Boulevard
Milpitas, CA 95035

MFM, INC.
360 Merrimack Street
Lawrence, MA 01843

MICROPOLIS CORPORATION
21123 Nordhoff Street
Chatsworth, CA 91311

MICROSCIENCE INTERNATIONAL CORPORATION
777 Palomar Avenue
Sunnyvale, CA 94086

MILTOPE CORPORATION
1770 Walt Whitman Road
Melville, NY 11747

MINISCRIBE CORPORATION
1871 Lefthand Circle
Longmont, CO 80501

NORTHERN TELECOM, INC.
Subsidiary of Northern Telecom, Ltd. (Canada)
259 Cumberland Bend
Nashville, TN 37228

PERIPHERAL TECHNOLOGY, INC.
685 East Cochran Street
Simi Valley, CA 93065

PERTEC PERIPHERALS CORPORATION
Subsidiary of Digital Development Corp
20400 Plummer Street
Chatsworth, CA 91311

PLUS DEVELOPMENT CORPORATION
Subsidiary of Quantum Corporation
1778 McCarthy Boulevard
Milpitas, CA 95035

PRIAM CORPORATION
20 West Montague Expressway
San Jose, CA 95134

QUANTUM CORPORATION
1804 McCarthy Boulevard
Milpitas, CA 95035

SEAGATE TECHNOLOGY
920 Disc Drive
Scotts Valley, CA 95066

STORAGE TECHNOLOGY CORPORATION
2270 South 88th Street
Louisville, CO 80027

SYQUEST TECHNOLOGY
47923 Warm Springs Boulevard
Fremont, CA 94538

TANDON CORPORATION
20320 Prairie Street
Chatsworth, CA 91311

TULIN CORPORATION
2393 Qume Drive
San Jose, CA 95131

UNISYS CORPORATION
Burroughs Place
Detroit, MI 48232

VERMONT RESEARCH CORPORATION
Precision Park
North Springfield, VT 05156

XEBEC
3579 Highway 50 East
Carson City, NV 89701

ASIAN MANUFACTURERS

(Firms are in Japan unless otherwise noted.)

ALPS ELECTRIC CO., LTD.
1–7, Yukigaya Otsuka-cho
Ohta-ku, Tokyo 145

COGITO SYSTEMS CORPORATION
Subsidiary of Ching Fong Investment Co., Ltd.
180 Chung Hsiao E. Rd., Sec. 4
Taipei, Taiwan

FUJI ELECTRIC CO., LTD.
12-1 Yurakucho 1-Chome
Chiyoda-ku
Tokyo, 100

FUJITSU LIMITED
6-1, Marunouchi 2-Chome
Chiyoda-ku, Tokyo 100

HITACHI, LTD.
4-6 Kanda-Surugadai
Chiyoda-ku, Tokyo 101

JVC (VICTOR COMPANY OF JAPAN, LTD.)
4-8 Nihonbashi-Honcho
Chuo-ku, Tokyo 103

MATSUSHITA COMMUNICATION INDUSTRIAL CO., LTD.
4-3-1 Tsunashima-Higashi
Kohoku-ku, Yokohama 223

MITSUBISHI ELECTRIC CORPORATION
2-3, Marunouchi 2-chome
Chiyoda-ku, Tokyo 100

NEC CORPORATION
5-33-1, Shiba
Minato-ku, Tokyo 108

NIPPON ELECTRIC INDUSTRY CO., LTD.
19-18, Tsutsumi-dori 1-Chome
Sumida-ku, Tokyo 131

NIPPON PERIPHERALS LIMITED
660 Miyamae, Fujisawa-shi
Kanagawa-ken 251

NIPPON SYSTEMHOUSE CO., LTD.
Nakajima Building
1-8-1, Kitashinjuku
Shinjuku-ku, Tokyo

OKI ELECTRIC INDUSTRY CO., LTD.
1-17-12, Toranomon
Minato-ku, Tokyo 105

ORIENTAL PRECISION COMPANY LIMITED
C.P.O. Box 1301
Seoul, Korea

OTARI ELECTRIC CO., LTD.
29-18, Minami Ogikubo 4-Chome
Suginami-ku, Tokyo 167

RICOH CO., LTD
15-5 Minami-Aoyama 1-Chome
Minato-ku, Tokyo 107

SEIKO EPSON CORPORATION
80 Hirooka
Shiojiri-shi, Nagano 399-07

SHINWA DIGITAL INDUSTRY CO., LTD.
1036 Kawarabuki
Ageo City, Saitama

SONY CORPORATION
6-7-35, Kita-Shinagawa
Shinagawa-ku, Tokyo 141

TATUNG CO.
22 Chungshan N. Road, Sec. 3
Taipei, Taiwan

TEAC CORPORATION
3-7-3, Naka-cho
Musashino, Tokyo 180

TOKICO, LTD
1-6-3, Fujimi
Kawasakiku, Kawasaki 210

TOKYO ELECTRIC CO., LTD
2-6, Naka-Meguro
Meguro-ku, Tokyo 153

TOSHIBA CORPORATION
1-1-1 Shibaura
Minato-ku, Tokyo 105

TOYO SODA MANUFACTURING CO., LTD.
1-7-7 Akasaka
Minato-ku, Tokyo 107

YE DATA, INC.
Subsidiary of Yaskawa Electric Mfg. Co., Ltd.
1-1 Higashi-Ikebukuro 3-Chome
Toshima-ku, Tokyo 170

EUROPEAN MANUFACTURERS

BASF AG
D-6700 Ludwigshafen
West Germany

BULL PERIPHERALS
94, Avenue Gambetta
75960 Paris Cedex 20
France

COMPAREX INFORMATIONSSYSTEME GMBH
Joint venture of BASF and Siemens
Gottlieb-Daimler-Strasse 10
D-6800 Mannheim
West Germany

ISOT
51, Chapaev St.
Sofia, bulgaria

KOVO
Jankovcova 2
17088 Praha 7
Czechoslovakia

LEXICON S.P.A. (Previously Olivetti Peripheral Equipment)
Subsidiary of Ing. C. Olivetti & C., S.p.A.
via Torina, 603
10090 S. Bernardo d'Ivrea (Torino)
Italy

NEWBURY DATA RECORDING, LTD
Subsidiary of Data Recording Instruments Co., Ltd.
Hawthorne Road, Staines
Middlesex TW18 3BJ
England

NIXDORF COMPUTER AG
Furstenallee 7
4790 Paderborn
West Germany

RODIME LIMITED
Nasmyth Road
Southfield Industrial Estates
Glenrothes, Fife KY6 2SD
Scotland

SAGEM
(Societe d'Applications Generales d'Electricite et de Mecanique)
6 Avenue d'Iena
75783 Paris CEDEX 16
France

SIEMENS AG
Communications Group
St. Martin-strasse 76
D-8000 Munchen 80
West Germany

Index

ABOUT THE AUTHOR

Hugh M. Sierra was an employee of International Business Machines for over 25 years, retiring in 1983. During that time he contributed greatly to the development of computer hardware, particularly to the technological foundations of direct access storage devices. In 1956 he helped design and test the Arithmetic Unit for the IBM RAMAC, which was the first computer equipped with a DASD. Mr. Sierra was later assigned to IBM's Advanced Technology Department, where in 1970 he helped design the Apollo Direct Access Storage Device (also known as Coronado and Whitney), which eventually became the IBM 3380. More recently, in 1980, he helped design and test IBM's Laser, Liquid-Crystal Display Device.

Mr. Sierra holds many U.S. and international patents, among them the patent for the Floating Point Arithmetic Control means for the IBM Model 650 computer and the Matrix Arithmetic System with input/output error checking circuits. The latter formed the basis for the IBM Model 1620 Computer Arithmetic Unit. He has written numerous internal IBM technical reports, and has also published in IEEE Proceedings, "Transactions on Electronic Computers" and "The Society for Information Display Digest."

Mr. Sierra is now a consultant practicing in Los Altos, California.